Lecture Notes in Information Systems and Organisation

Volume 16

More information about this series at http://www.springer.com/series/11237

Fred D. Davis • René Riedl • Jan vom Brocke •
Pierre-Majorique Léger • Adriane B. Randolph
Editors

Information Systems and Neuroscience

Gmunden Retreat on NeuroIS 2016

Editors
Fred D. Davis
Informaton Systems
Texas Tech University
Lubbock, Texas
USA

René Riedl
University of Appl. Sc. Upper Austria
Steyr, Oberösterreich
Austria

University of Linz
Linz, Austria

Jan vom Brocke
Department of Information Systems
University of Liechtenstein
Vaduz, Liechtenstein

Pierre-Majorique Léger
Department of Information Technology
HEC Montréal
Montréal, Québec
Canada

Adriane B. Randolph
Department of Information Systems
Kennesaw State University
Kennesaw, Georgia
USA

ISSN 2195-4968 ISSN 2195-4976 (electronic)
Lecture Notes in Information Systems and Organisation
ISBN 978-3-319-41401-0 ISBN 978-3-319-41402-7 (eBook)
DOI 10.1007/978-3-319-41402-7

Library of Congress Control Number: 2016952620

Printed on acid-free paper

This Springer imprint is published by Springer Nature
The registered company is Springer International Publishing AG Switzerland

Preface

NeuroIS is a field in Information Systems (IS) that makes use of neuroscience and neurophysiological tools and knowledge to better understand the development, adoption, and impact of information and communication technologies. The *Gmunden Retreat on NeuroIS* is a leading academic conference for presenting research and development projects at the nexus of IS and neurobiology (see http://www.neurois.org/). This annual conference has the objective to promote the successful development of the NeuroIS field. The conference activities are primarily delivered by and for academics, though works often have a professional orientation.

The conference is taking place in Gmunden, Austria, a much frequented health and summer resort providing an inspiring environment for the retreat. In 2009, the inaugural conference was organized. Established on an annual basis, further conferences took place from 2010 to 2015.

The genesis of NeuroIS took place in 2007. Since then, the NeuroIS community has grown steadily. Scholars are looking for academic platforms to exchange their ideas and discuss their studies. The *Gmunden Retreat on NeuroIS* seeks to stimulate these discussions. The conference is best characterized by its "workshop atmosphere." Specifically, the organizing committee welcomes not only completed research but also work in progress. A major goal is to provide feedback for scholars to advance research papers, which then, ultimately, have the potential to result in high-quality journal publications.

This year is the second time that we publish the proceedings in the form of an edited volume. A total of 24 research papers are published in this volume, and we observe diversity in topics, theories, methods, and tools of the contributions in this book. Altogether, we are happy to see the ongoing progress in the NeuroIS field.

Increasingly more people, both IS researchers and practitioners, have been recognizing the enormous potential of neuroscience tools and knowledge.

Lubbock, TX Fred D. Davis
Steyr, Austria René Riedl
Vaduz, Liechtenstein Jan vom Brocke
Montréal, Canada Pierre-Majorique Léger
Kennesaw, GA Adriane B. Randolph

May 2016

Contents

Tracing Consumers' Decision-Making in Digital Social Shopping Networks

Camille Grange and Izak Benbasat

Abstract Several online businesses consider integrating social network functionalities into their online store, thus creating digital social shopping networks (DSNs). Because friends tend to be trusted or have similar interests, one often expects that shoppers will leverage their connections to simplify choice making and enhance the chances of finding well-suited products. Unfortunately, there is little evidence or theory to justify whether, when, and how this might manifest. In this research, we address this issue by developing a model of how decision-making unfolds in a DSN. The model can be used as a basis for a research program aimed at explaining why, when, and by what means online shoppers leverage their social network or cease leveraging it. We outline the key features of an empirical strategy, which involves process tracing via the use of complementary techniques: navigation recording, verbal protocol and ocular activity.

Keywords Online shopping • Social networks • Cybernetic theory • Process tracing • Verbal protocols • Eye-tracking

1 Introduction

With worldwide sales reaching $1.5 trillion this year, business-to-consumer e-commerce is more prevalent than ever and continues to be a powerful driver of retail growth [1]. An important challenge associated with this phenomenon is that it yields almost limitless product assortments and an overwhelming amount of consumer-generated content (e.g., ratings and reviews), making online shoppers' decision-making process complex and often dissatisfying [2]. For example, searching for digital cameras on Amazon returns more than 40,000 alternatives,

C. Grange (✉)
HEC Montreal, Montreal, QC, Canada
e-mail: camille.grange@hec.ca

I. Benbasat
University of British Columbia, Vancouver, QC, Canada
e-mail: izak.benbasat@sauder.ubc.ca

F.D. Davis et al. (eds.), *Information Systems and Neuroscience*, Lecture Notes in Information Systems and Organisation 16, DOI 10.1007/978-3-319-41402-7_1

and searching for peers' opinions on Thailand on TripAdvisor reports more than three million reviews.

Designing online stores in such a way that shoppers are able to connect to and access opinions from friends could help mitigate this issue because friends can make trustworthy and relevant recommendations, filtering out the informational clutter from which contemporary online environments suffer. The potential of such information technologies, which we refer to as *digital social shopping networks* (DSNs), is thus promising. In fact, the results of a recent study suggest that DSNs can enhance the product search experience by yielding more serendipity [3], i.e., unexpected but useful discoveries. In that study, the mechanisms underlying the emergence of serendipity relied on *how* socially embedded shoppers navigate the product space, which points to the utility of focusing on shoppers' decision-making *process* as a unit of analysis to examine the potential value of DSNs.

Therefore, our aim in the present research is to examine socially embedded shoppers' decision-making process, focusing on finding out when, why, and how they initiate, engage, and cease activating their social network to identify and evaluate products of interest. We leverage the decision-making and online shopper behavior literatures to devise a research model that offers a contingent cybernetic view of consumers' decision-making process in a DSN setting. The model is *cybernetic* in that it assumes that online shoppers self-regulate their search behavior by feedback loops. In other words, the model proposes that shoppers engage in and revise a set of product search tactics in order to meet their goals (e.g., learning, being efficient, finding a new product). The model is *contingent* in that it assumes that shoppers may pursue different strategies (e.g., social vs. non-social) depending on the potential value they attribute to their network.

This research contributes to the literatures on consumer decision-making and digital networks. Regarding the former, substantial research has reported that consumers are pragmatic beings who seek satisficing ways to reach their goals [4], and that their decision-making process reflects a contingent form of information processing [5] that involves a degree of self-regulation [6]. We intend to add to this literature by explaining how these assumptions apply in online social environments. Regarding the latter, there has been a growing interest in digital network research [e.g., 7–15]. Yet, very little work has yet examined what such networks imply for e-commerce in particular—see [16] and [17] for exceptions. We extend this nascent literature by shedding light into how socially embedded shoppers' product search unfolds and what may prevent, constrain, or enable its success.

The next section presents the study background and the research model. The third section outlines our anticipated empirical strategy. Finally, we articulate what we feel are the most original and interesting potential contributions of this research.

2 Theoretical Development

The kind of search behavior studied in this research is a typical pre-purchase activity [18], which consists in shoppers conducting a product selection task within a particular category (e.g., movies), for which they need to navigate throughout a website to identify and evaluate alternatives that may fit their needs and interests. To examine this phenomenon, we adopt a contingent perspective that applies the principles of cybernetic theory [19], a framework used in several studies of human and organizational behavior in information systems or related fields [6, 20]. In the context of online consumer decision-making, this view implies that shoppers engage in a self-regulatory sequence of activities that involve (1) detecting what the decision-making environment affords (***perception***), (2) comparing perceptions to search goals (***comparison***), (3) leveraging the affordances of the environment to pursue product and information search (***actions***), and (4) assessing whether the action was successful and if more search is needed (***consequence***). The model also acknowledges that ***disturbances*** can happen, that is, there are risks that a shopper might misdirect efforts (i.e., overinvest in search which will not yield additional insight or underinvest and thus not be able to obtain enough information to make a good choice). Figure 1 represents our application of this view to the study of consumer decision-making in online settings that afford access to friends and their product evaluations.

Thus, the proposed model submits the following contingent loop of activities:

1. First, shoppers initiate a sense-making process about what the platform they interact with affords (***SN consideration***). Two aspects can be distinguished here: *sensory* (mostly visual, in our context) and *cognitive* (understanding). Similar to others [e.g., 21], we consider a website as a stimuli-based decision-making environment. In a DSN context, stimuli take the form of features such as links to access a list of one's direct SN contact (e.g., list of friends) or a list of these contacts' opinions towards products (e.g., social feed). We anticipate the

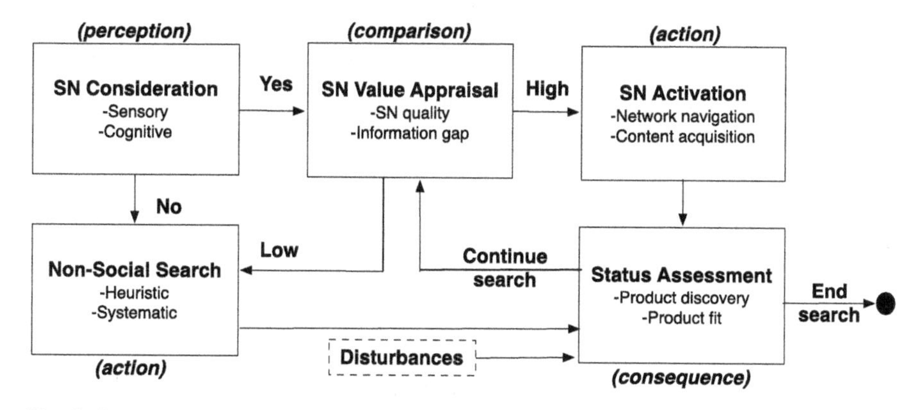

Fig. 1 Proposed theoretical model

salience of those stimuli to a shopper to depend on the website's design but also on the idiosyncratic properties of her network (e.g., how many friends she has, how active they have been in creating content).

2. Second, shoppers assess the potential value of using their network for pursing the product selection task (***SN value appraisal***). A framework particularly useful to think about SN value appraisal is information foraging, a theory that focuses on the question of how individuals navigate information spaces such as websites [22, 23]. An important concept of this theory is information scent; it refers to the degree of informativeness of the stimuli or cues (e.g., filters and links on websites) noticed and eventually used by information foragers to select which information sources to pursue and consume [24, 25]. According to this theory, information seekers adjust navigation behaviors to follow high-scent cues so as to ultimately maximize the rate of gaining valuable information, i.e., be more efficient with respect to their information search [22]. In line with this view, we expect shoppers to evaluate their network's potential value based on two aspects: (i) the ability of friends in their network to offer reliable and interesting product recommendation—i.e., social *network quality* criteria such as how trustworthy friends are [26], and (ii) their own curiosity in filling an *information gap* by discovering what familiar others (friends) have bought or recommended [27].
3. When SN value is sufficiently[1] high, shoppers should be motivated to exploit their network (***SN activation***). In line with the traditional two-stage view of decision-making (the screening of product alternatives followed by the in-depth evaluation of a smaller consideration set), we conceptualize SN activation via two kinds of actions: (i) *navigating* the network, an information-seeking behavior that involves relying on one's friends to reach certain products of interest [28], and (ii) *acquiring content* in the network, i.e., obtaining the opinions or advice from friends on products of interest.
4. ***Non-social search*** refers to shoppers' actions that do not involve using their network. Such actions may include the use of other *heuristics* such as considering or choosing the cheapest or the most popular products only, or the pursuit of a more *systematic* strategy that relies on an elaborate, logical, product selection and evaluation process. We expect shoppers to engage in non-social product search when they have either not noticed the social signals conveyed by the DSN or when the perceived informational scent of their SN does not match the information requirements dictated by their goal (e.g., when they believe that friends are not relevant enough to feel the need to fill an information gap or are not able to make a good recommendation).
5. Last, the consequence stage is reflected in the ***status assessment*** construct, defined as shoppers' evaluation of whether they should keep on searching for additional product alternatives or information about them. One reason for

[1]The term sufficient has a pragmatic connotation that fits with the idea that SN value is assessed by users based on their own standards and needs.

deciding to exit the process is the feeling of being able to make a sufficiently good decision considering the additional effort that may be required to keep on identifying or evaluating alternatives. While different shoppers may prioritize different criteria for assessing the status of their search, we expect that *discovery* [29] and/or *fit* [30] will be important ones. Another (undesirable) reason for exiting the process may be the feeling that the environment does not and will not be able to help the shopper make up her mind.

3 Empirical Strategy

We will use an approach that enables capturing behavioral, cognitive, and neuro-physiological measures in a controlled environment. Experimental subjects will be asked to use a custom-developed DSN to conduct a product (mobile apps and movies) selection task. The concepts and relationships presented in Sect. 2 will guide our empirical enquiry. Because the research model specifies constructs of different nature (attention, cognition, website navigation), we plan on collecting data via a mix of process tracing methods [31, 32].

The experiment will employ a 2×2 factorial design, with product type (movie vs. mobile app) and time constraint (low vs. high) as the two factors. These two factors were chosen to create variance in shopping contexts and enhance external validity. Our custom-built experimental platform will be populated with real instances of movies and mobile apps, and will connect to the Facebook's API to obtain our study participants' social network data, which we will use to create functionalities that afford accessing friends and their product recommendations, as we explain next.

Our data collection strategy will unfold in two stages. The first stage will happen exclusively online. We will recruit an initial set of participants using a panel at our university. Once they register (using their Facebook credentials), they will complete a review task, i.e., rate and give their opinions on the apps and movies they know of, and they will be incentivized to refer our study to their Facebook friends; we will also use Facebook to advertise the study further to a specific audience restricted to the Facebook friends of those already signed up. Those mechanisms (referral and targeted advertising) will be used to ensure that we obtain a sample of interconnected participants (without which we would not be able to assess our model). Second, approximately 1 month after having completed the first stage, we will get in touch again with the participants who completed the first stage, and invite them to come to our lab for a follow-up study, during which they will conduct a product choice task and the data relevant to our research model will be collected.

Returning subjects will be randomly assigned to five groups (four groups determined by our factorial design, and a control group in which the experimental website will not connect subjects to their social network). In the low time constraint condition, subjects will be given no time limit while in the high time constraint, a 5 min countdown timer will be displayed on the screen. Subjects will be asked to browse the site to compile a list of three apps or movies they would be interested in purchasing. They will receive money rebates on selected products. This 'product discovery' task aims at motivating subjects to search for information that will assist them in making a satisfying decision. During this product search task, subjects will be connected to their 'natural' social network. In other words, the review content created by their Facebook friends who participated to the study will be made accessible to them via 'list of friends' and 'feed of friends' recommendations' features.

During this experiment, we will collect a set of behavioral, cognitive, and neurophysiological data. As for behaviors, users' mouse movements will be automatically recorded via asynchronous JavaScript coded into the website and a Camtasia video recorder. Ocular activity (fixations and saccades) and verbal protocols are two additional methods that complement each other in supporting the tracing of experimental subjects' information processing [33]. Thus, we plan to use concurrent neutral-probing verbal protocol, which is considered a powerful method for discovering the cognitive dynamics of information search and processing in unstructured environments [34]. A small portion of our subjects (5–15 range) will be asked to "think aloud" while simultaneously engaging in the task, their computer screen being video-recorded. Autonomic nervous system activity will be captured via a Tobii screen-based eye tracker, to record eye fixations and movements on the totality of our sample. Fixation patterns refer to short gazing stops where the eye is able to process information. This data will provide insights on the subjects' focus of attention during the completion of their task, thus helping in particular to capture aspects related to 'SN consideration' (Fig. 1). Eye fixations and movements tend to reflect though processes, so they will offer a trace of the choice process that is not easily censored by subjects, and thus it will serve as a useful supplement to the verbal protocols and navigation data by indicating which information is attended to and used irrespective of what subjects say or do [35]. To analyse this data (navigation pattern, verbal protocols, eye movements). Two people from our research team will tabulate the frequency of occurrence of items of interest using a coding scheme that will be developed a priori based on the research model's constructs and the four key aspects characterizing an information search strategy (content type, depth, sequence, latency) [36].

4 Expected Research Contributions and Extensions

This research is a work-in-progress, and thus a first step toward developing a better understanding of online shoppers' product search experience when using a digital social shopping network. Ultimately, we hope the knowledge gained from this research will help UX practitioners to design environments that better support online shoppers' information search and decision-making. While prior research found that decision-makers adapt to their informational environment [37] and seek to maximize value while minimizing effort [38], we aim to contribute further by explaining if and how such assumptions manifest in the case of socially-embedded shoppers. In addition, the proposed process perspective leaves some room for exploration and should reveal a greater variety of potential outcomes compared to a more traditional 'input-output' approach (i.e., effect of X on Y) [39], thus offering useful implications for future work such as finding out which variables—perhaps unexpected ones—are worth studying in greater depth.

Other neurophysiological methods offer opportunities to expand this research by enabling to capture more 'immediate' effect of DSNs on consumer behaviors, cognitions and emotions [40–44]. The model we presented in this paper is situated at a high-level of abstraction, leaving the room for more micro-level investigations that would focus on specific facets of shoppers' use of DSNs. For example, much needs to be learned about shoppers' emotional reactions (e.g., arousal) to the presence of friends in online shopping environments, and techniques such as electromyography (EMG), electro-dermal activity (EDA), electroencephalography (EEG), and functional magnetic resonance imaging (fMRI) would be useful in that regard [44]. Trust in friends is another key construct in the study of decision-making in DSNs, and techniques such as fMRI could help inform researchers about the potential of DSN at reducing uncertainty by inducing a sense of credibility and/or benevolence from advice givers present in one's network [45–47]. Finally, because mental workload is a constant concern in the design of online shopping environments, the use of eye-tracking-based pupillometry could help investigate whether SNs can contribute to decrease effort or become more useful in high workload situations [48–50].

References

1. eMarketer: Global B2C Ecommerce Sales to Hit $1.5 Trillion This Year Driven by Growth in Emerging Markets (2014)
2. Iyengar, S.S., Lepper, M.R.: When choice is demotivating: can one desire too much of a good thing? J. Pers. Soc. Psychol. **79**, 995–1006 (2000)
3. Grange, C.: Three essays on the value of online social commerce. https://circle.ubc.ca/handle/2429/48386 (2014)
4. Gigerenzer, G., Gaissmaier, W.: Heuristic decision making. Annu. Rev. Psychol. **62**, 451–482 (2011)
5. Payne, J.W.: Contingent decision behavior. Psychol. Bull. **92**, 382–402 (1982)
6. Byrnes, J.P.: The Nature and Development of Decision-making: A Self-regulation Model. Psychology Press, London (2013)
7. Butler, B., Matook, S.: Social media and relationships. In: Mansel, R. (ed.) The International Encyclopaedia of Digital Communication and Society, pp. 1–20. Wiley-Blackwell, London (2014)
8. Ellison, N., Boyd, D.: Sociality through social network sites. In: Dutton, W.H. (ed.) The Oxford Handbook of Internet Studies. Oxford University Press, Oxford (2013)
9. Hildebrand, C., Häubl, G., Herrmann, A., Landwehr, J.R.: When social media can be bad for you: community feedback stifles consumer creativity and reduces satisfaction with self-designed products. Inf. Syst. Res. **24**, 14–29 (2013)

10. Kane, G.C., Alavi, M., Labianca, J., Borgatti, S.P.: What's different about social media networks? A framework and research agenda. MIS Q. **38**, 274–304 (2014)
11. Krasnova, H., Widjaja, T., Buxmann, P., Wenninger, H., Benbasat, I.: Research note—why following friends can hurt you: an exploratory investigation of the effects of envy on social networking sites among college-age users. Inf. Syst. Res. **26**, 585–605 (2015)
12. Majchrzak, A., Faraj, S., Kane, G.C., Azad, B.: The contradictory influence of social media affordances on online communal knowledge sharing. J. Comput.-Mediat. Commun. **19**, 38–55 (2013)
13. Sundararajan, A., Provost, F., Oestreicher-Singer, G., Aral, S.: Research commentary—information in digital, economic, and social networks. Inf. Syst. Res. **24**, 883–905 (2013)
14. Toubia, O., Stephen, A.T.: Intrinsic vs. image-related utility in social media: why do people contribute content to Twitter? Mark. Sci. **32**, 368–392 (2013)
15. Zeng, X., Wei, L.: Social ties and user content generation: evidence from Flickr. Inf. Syst. Res. **24**, 71–87 (2013)
16. Wang, C., Zhang, P.: The evolution of social commerce: the people, management, technology, and information dimensions. Commun. Assoc. Inf. Syst. **31**, 105–127 (2012)
17. Grange, C., Benbasat, I.: Foundations for investigating the drivers of the value captured by consumers embedded within social shopping networks. In: Proceedings of the 46th Hawaii International Conference on System Sciences (HICSS), pp. 680–689. Maui, HI (2013)
18. Bloch, P.H., Sherrell, D.L., Ridgway, N.M.: Consumer search: an extended framework. J. Consum. Res. **13**, 119–126 (1986)
19. Wiener, N.: Cybernetics: Control and Communication in the Animal and the Machine. MIT Press, Cambridge, MA (1948)
20. Edwards, J.R.: A cybernetic theory of stress, coping, and well-being in organizations. Acad. Manage. Rev. **17**, 238–274 (1992)
21. Tam, K.Y., Ho, S.Y.: Understanding the impact of web personalization on user information processing and decision outcomes. MIS Q. **30**, 865–890 (2006)
22. Pirolli, P., Card, S.: Information foraging. Psychol. Rev. **106**, 643–675 (1999)
23. Pirolli, P.L.: Information Foraging Theory: Adaptive Interaction with Information. Oxford University Press, Oxford (2007)
24. Chi, E.H., Pirolli, P., Chen, K., Pitkow, J.: Using information scent to model user information needs and actions and the web. In: Proceedings of the SIGCHI Conference on Human Factors in Computing Systems, pp. 490–497. ACM, New York (2001)
25. Pirolli, P.: The use of proximal information scent to forage for distal content on the world wide web. In: Kirlik, A. (ed.) Adaptive Perspectives on Human-Technology Interaction: Methods and Models for Cognitive Engineering and Human-Computer Interaction, pp. 247–266. Oxford University Press, Oxford (2006)
26. Bonhard, P., Harries, C., McCarthy, J., Sasse, M.: Accounting for taste: using profile similarity to improve recommender systems. In: Proceedings of the Conference on Human Factors in Computing Systems (CHI), pp. 1057–1066. ACM, Montreal (2006)
27. Loewenstein, G.: The psychology of curiosity: a review and reinterpretation. Psychol. Bull. **116**, 75–98 (1994)
28. Munro, A.J., Hook, K., Benyon, D.: Social Navigation of Information Space. Springer, London (1999)
29. Herlocker, J.L., Konstan, J.A., Terveen, L.G., Riedl, J.T.: Evaluating collaborative filtering recommender systems. ACM Trans. Inf. Syst. **22**, 5–53 (2004)
30. Smith, S.P., Johnston, R.B., Howard, S.: Putting yourself in the picture: an evaluation of virtual model technology as an online shopping tool. Inf. Syst. Res. **22**, 640–659 (2011)
31. Payne, J.W.: Task complexity and contingent processing in decision making: an information search and protocol analysis. Organ. Behav. Hum. Perform. **16**, 366–387 (1976)
32. Russo, J.E., Dosher, B.A.: Strategies for multiattribute binary choice. J. Exp. Psychol. Learn. Mem. Cogn. **9**, 676–696 (1983)

33. Russo, J.E.: Eye fixations can save the world: a critical evaluation and a comparison between eye fixations and other information processing methodologies. Adv. Consum. Res. **5**, 561–570 (1978)
34. Ericsson, K.A., Simon, H.A.: Protocol Analysis: Verbal Reports as Data. MIT Press, Cambridge, MA (1993)
35. Léger, P.-M., Sénecal, S., Courtemanche, F., de Guinea, A.O., Titah, R., Fredette, M., Labonte-LeMoyne, É.: Precision is in the eye of the beholder: application of eye fixation-related potentials to information systems research. J. Assoc. Inf. Syst. **15**, 651–678 (2014)
36. Ford, J.K., Schmitt, N., Schechtman, S.L., Hults, B.M., Doherty, M.L.: Process tracing methods: contributions, problems, and neglected research questions. Organ. Behav. Hum. Decis. Process. **43**, 75–117 (1989)
37. Payne, J.W., Bettman, J., Johnson, E.J.: The Adaptive Decision-Maker. Cambridge University Press, New York (1993)
38. Johnson, E.J., Payne, J.W.: Effort and accuracy in choice. Manag. Sci. **31**, 395–414 (1985)
39. Todd, P., Benbasat, I.: Process tracing methods in decision support systems research: exploring the black box. MIS Q. **11**, 493–512 (1987)
40. Dimoka, A., Banker, R.D., Benbasat, I., Davis, F.D., Dennis, A.R., Gefen, D., Gupta, A., Ischebeck, A., Kenning, P.H., Pavlou, P.A., Müller-Putz, G., Riedl, R., vom Brocke, J., Weber, B.: On the use of neurophysiological tools in IS research: developing a research agenda for NeuroIS. MIS Q. **36**, 679–702 (2012)
41. Dimoka, A., Pavlou, P., Davis, F.: Research commentary—NeuroIS: the potential of cognitive neuroscience for information systems research. Inf. Syst. Res. **22**, 687–702 (2011)
42. Riedl, R., Davis, F., Hevner, A.: Towards a NeuroIS research methodology: intensifying the discussion on methods, tools, and measurement. J. Assoc. Inf. Syst. **15** (2014)
43. Ortiz de Guinea, A., Webster, J.: An investigation of information systems use patterns: technological events as triggers, the effects of time, and consequences for performance. MIS Q. **37**, 1165–1188 (2013)
44. Gregor, S., Lin, A.C.H., Gedeon, T., Riaz, A., Zhu, D.: Neuroscience and a nomological network for the understanding and assessment of emotions in information systems research. J. Manag. Inf. Syst. **30**, 13–48 (2014)
45. Dimoka, A.: What does the brain tell us about trust and distrust? Evidence from a functional neuroimaging study. MIS Q. **34**, 373–396 (2010)
46. Dimoka, A., Hong, Y., Pavlou, P.A.: On product uncertainty in online markets: theory and evidence. MIS Q. **36**, 395–426 (2012)
47. Riedl, R., Mohr, P.N.C., Kenning, P.H., Davis, F.D., Heekeren, H.R.: Trusting humans and avatars: a brain imaging study based on evolution theory. J. Manag. Inf. Syst. **30**, 83–114 (2014)
48. Buettner, R.: Investigation of the relationship between visual website complexity and users' mental workload: a NeuroIS perspective. In: Davis, F.D., Riedl, R., vom Brocke, J., Léger, P.-M., Randolph, A.B. (eds.) Information Systems and Neuroscience, pp. 123–128. Springer, Switzerland (2015)
49. Buettner, R., Daxenberger, B., Eckhardt, A., Maier, C.: Cognitive workload induced by information systems: introducing an objective way of measuring based on pupillary diameter responses. SIGHCI 2013 Proceedings (2013)
50. Wang, Q., Yang, S., Liu, M., Cao, Z., Ma, Q.: An eye-tracking study of website complexity from cognitive load perspective. Decis. Support Syst. **62**, 1–10 (2014)

Validating Knowledge Filtering Processes in Electronic Networks of Practice

Thomas O. Meservy, Kelly J. Fadel, Rayman D. Meservy, Christopher Doxey, and C. Brock Kirwan

Abstract Online forums sponsored by Electronic Networks of Practice (ENPs) have become an important source of information for individuals seeking solutions to problems online. However, not all information available in a forum is helpful or accurate, requiring knowledge seekers to evaluate and filter the solutions they encounter. Most forums offer peripheral cues to help knowledge seekers make evaluation decisions, yet little is understood about the information filtering patterns people engage in when filtering information on ENP forums. This paper reports the results of a pilot fMRI study designed to elucidate the neural correlates involved in evaluating both content and peripheral cues encountered on an ENP forum. The ultimate goal of this research is to better understand information processing patterns in online forums and how these patterns produce more or less accurate filtering results.

Keywords fMRI • Electronic network of practice • Information filtering • Online forum • Programming experiment

1 Introduction and Theory

Electronic networks of practice (ENPs) are collections of geographically separated individuals who share common interests but are loosely affiliated and communicate via technology-mediated channels [1]. One of the most common tools used by ENPs is the online knowledge forum, a virtual bulletin board where ENP members can post and retrieve knowledge about various problems relevant to network members. For example, a programmer searching for a solution to a programming problem might access an online forum such as stackoverflow.com or dreamincode.net to post a query requesting information from other ENP members or to search for

T.O. Meservy (✉) • R.D. Meservy • C. Doxey • C.B. Kirwan
Brigham Young University, Provo, UT, USA
e-mail: tmeservy@byu.edu; meservy@byu.edu; doxeyc@byu.edu; kirwan@byu.edu

K.J. Fadel
Utah State University, Logan, UT, USA
e-mail: kelly.fadel@usu.edu

F.D. Davis et al. (eds.), *Information Systems and Neuroscience*, Lecture Notes in Information Systems and Organisation 16, DOI 10.1007/978-3-319-41402-7_2

existing queries that match her own. In response, ENP members post potential solutions, often in the form of actual code blocks. Surrounding these solutions, most forums offer additional peripheral cues to the information seeker, such as characteristics of the respondent (e.g., expertise level), endorsement of the solution by an expert ENP member, or validation of the solution by other members of the community.

To-date, research has explored what motivates individuals to contribute knowledge to an ENP forum [1, 2]; however, research that examines how individuals evaluate and ultimately adopt knowledge from such forums is more scarce. Recent studies in this area have employed dual process theories of cognition [e.g., 3], which suggest that, depending on their expertise and motivation, knowledge seekers may engage both central route processing (deliberate evaluation of the knowledge content itself) and peripheral route processing (interpretation of various peripheral cues surrounding the knowledge) when evaluating information. Results show that even people who possess the expertise to evaluate information on its own merits often rely on peripheral cues available in ENP forums (such as third-party validation of a solution) in deciding what information to adopt [4]. Moreover, attentional switching patterns between central and peripheral processing employed by knowledge seekers also appears to influence the normative accuracy of filtering decisions [5].

While the studies cited above have begun to yield insight into the ways that people evaluate information encountered in ENP forums, they are limited in their ability to elucidate the cognitive processes that underlie filtering decisions. As with many behavioral IS studies examining similar phenomena, cognitive processing is inferred through either observed behavioral patterns or through gaze data captured by an eye-tracking device. Advances in neuroimaging technologies and their application in the IS field have created new opportunities to "further advance our knowledge of the complex interplay of IT and information processing, decision making, and behavior" [6, p. 687]. Our objective in this study is to leverage fMRI data to localize brain areas responsible for both central and peripheral route processing of information encountered on an ENP forum, thereby achieving more complete understanding of the cognitive mechanisms underlying information filtering behaviors.

Although the application of neuroimaging to dual process cognition is, to our knowledge, novel in IS literature, some studies have shown distinct neurological patterns associated with peripheral and central route processing in other domains. For example, Goel et al. [7, 8] found that in syllogistic reasoning tasks, stimuli that conformed to a belief bias (a heuristic or peripheral route processing strategy) activated a different area of the brain than those that were not amenable to this heuristic and therefore required more concentrated logical analysis. Specifically, the right lateral prefrontal cortex was associated with more in-depth central-route thinking, while the ventral medial prefrontal cortex was more active in heuristic-based judgments. Later studies have offered further evidence for this distinction [9, 10]. A meta-analysis of previous studies reporting activations in these regions indicates that activations in the right lateral prefrontal cortex is most commonly associated with the terms "beliefs" and "reappraisal" [11]. Activation in the ventromedial prefrontal cortex is most commonly associated with "theory of mind" [12]. Accordingly, it seems that central processing may be associated with updating or reappraising one's own beliefs while peripheral processing may be associated with evaluating the beliefs of others. Based on the results of this research, we hypothesize the following:

H1: Central route processing (evaluation of the solution content itself) is associated with higher activation in the dorsolateral prefrontal cortex than peripheral route processing (evaluation based on peripheral cues).

H2: Peripheral route processing exhibits higher activation in the ventromedial prefrontal cortex than central-route processing.

2 Methodology

We used functional magnetic resonance imaging (fMRI) to observe the cognitive processing patterns of software developers as they were presented and asked to rate solutions to problems that were found online. Six experienced software developers (all male, average age 29 years, average of 4.8 years programming experience) were recruited from local companies to participate in the pilot study and were compensated with $25 or a 3D model of their brain for their participation. Participants were first shown an introductory video to acclimate them to the experimental instrument and task. After a localization and structural scan in the MRI, participants were trained in the experimental instrument. They were then presented with two separate blocks of questions/solutions that are described below. After the experiment, a short survey was used to gather demographics and perceptions about the experimental task.

During the experiment, participants were shown a total of 48 programming solutions in sequence (eight solutions for each of six programming problems). Problems were selected that were easily understandable and could be evaluated in a short amount of time but that still required a systematic evaluation to assess the quality of the proposed solution (e.g., sorting an array, testing whether a word is a palindrome). The experimental interface was modeled after actual online programming forums, and showed the code that purportedly would solve the problem as well as two peripheral cues, expert rating and community rating, that attested to the validity (or lack thereof) of the presented solution. Additional information such as the name of the expert who provided the judgment, an avatar, join date, and number of posts were also included to increase the face validity of the interface. Solutions varied based on their correctness (actually solves the problem or doesn't), expert rating (recommended or not) and community rating (large percentage recommended it or did not recommend it). Solutions for a given problem were presented in sequence to minimize cognitive burden; however, all other aspects of stimulus presentation were randomized, including problem order, solution order within each problem, expert ratings, community ratings, and

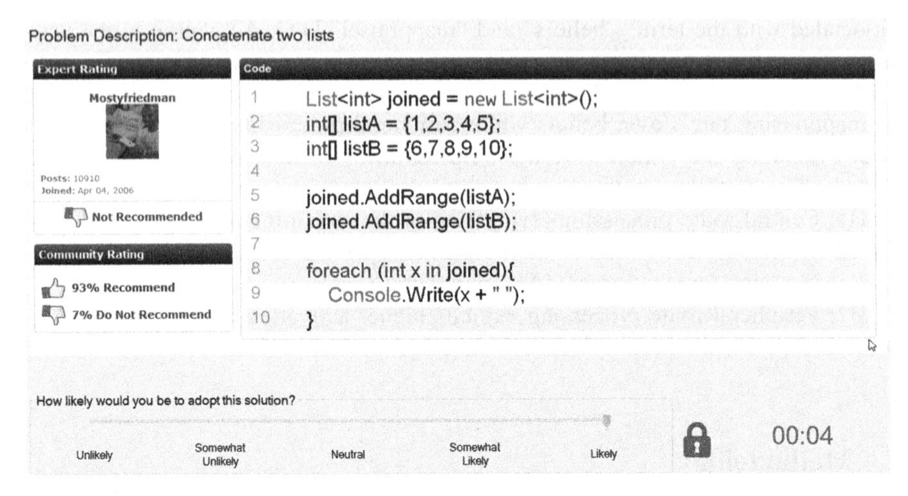

Fig. 1 Sample stimulus

information about the expert who provided the rating. Expert names and images were extracted from actual ENP forums. Figure 1 shows a solution for one of the problems presented.

Participants viewed the experimental instrument on a large MR-compatible monitor at the opening of the MRI scanner by means of a mirror attached to the head coil. Participants used a 4-button controller to interact with the experimental instrument and provide ratings of their likelihood of adopting the presented solutions. To isolate cognitive processing differences in peripheral processing (evaluation based on expert and community ratings) vs. central processing (evaluation of the solution itself), we controlled the visibility of solution components. Because peripheral cues are often used to filter potential solutions [4], participants were first shown only the peripheral cues with the code blurred and were required to rate their likelihood of adopting the solution based on these cues alone. After entering their rating, the code was un-blurred and participants were allowed to adjust their rating if desired after examining the code itself. Each solution was presented for a total of 30 seconds.

3 Analysis and Results

Each participant in the pilot study contributed a standard-resolution structural scan, and two functional MRI (fMRI) scans using a 3T Siemens Tim Trio scanner. During the fMRI scans, participants completed the behavioral task. Imaging data were analyzed using the Analysis of Functional Neuroimages (AFNI) suite of programs [13] and the Advanced Normalization Tools (ANTs) [14, 15] following common procedures including co-registration, correcting head motion, and

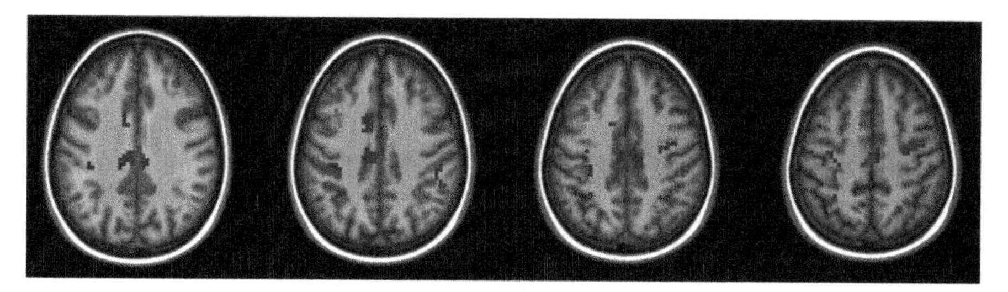

Fig. 2 Activation based on evaluation of central vs. peripheral cues

spatially normalized for group analyses. Behavioral vectors coded for trials where participants were either examining the peripheral cues or both peripheral and central cues were created in the first level regression analysis. The fMRI model also included vectors that coded for scan run, scanner drift, and motion. The resulting beta coefficients were entered into group-level analyses as described below. We corrected for multiple comparisons using the AFNI ClustSim program, which uses Monte Carlo simulations to calculate the appropriate clusters of voxels that are large enough to be statistically significant [16, 17]. Using a voxel-wise threshold p-value of <0.001 the calculated minimum cluster threshold was 20 voxels.

Using the deconvolved fMRI data, we conducted pairwise t-tests to evaluate differences in brain activity for central cues vs. peripheral cues. The analysis revealed 14 significant activation clusters. Three clusters exhibited more activity for central processing (evaluating the solution code itself) including left middle frontal gyrus, left inferior frontal gyrus, and left superior frontal gyrus. Two of the clusters that produced higher levels of activation when evaluating code are part of the dorsolateral prefrontal cortex and the third cluster is part of the ventrolateral prefrontal cortex. These results provide support for H1.

Figure 2 highlights brain regions that are more active for central processing (orange) and peripheral processing (blue).

4 Discussion and Conclusion

Significant activation related to central route processing supports the idea that evaluating code solutions on an ENP forum requires higher cognitive functions. For example, the superior frontal gyrus contributes to higher cognitive functions and specifically to working memory [18] and has been unequivocally linked to switching between distinct cognitive tasks [19], whereas the left inferior frontal gyrus is involved in the control of information from long-term memory [20] and has consistently been associated with semantic selection, processing and operations [21, 22]. The functionality provided by these areas is clearly important when evaluating blocks of code and assessing their likelihood of solving a specific problem.

When evaluating the peripheral cues, participants took into account expert and community ratings that at times were conflicting in order to arrive at a judgment of the quality of the solution. We suspect that several of the significant differences related to peripheral cues appear due to increased levels of hand movement that stem from selecting or changing ratings. Previous studies have reported that many aspects for the parietal lobe are directly related to motor functions including sensory input from hands (superior parietal lobule), sense of touch (postcentral gyrus), and interpretation of sensory information (inferior parietal lobule) [23, 24]. Results related to the anterior and posterior cingulate merit further investigation. While these areas have been implicated in visceromotor, skeletomotor and eye movements [25], they have also been shown to be linked to rational cognition including rewards, decision making, and cognitive control [26, 27].

This study is among the first to provide empirical evidence of the diverse cognitive processes involved in to filtering information in online ENP forums. Our goal is to build on these results in future studies to explore how different types of cognitive processing affect the accuracy of information filtering decisions.

References

1. Wasko, M.M., Faraj, S.: Why should I share? Examining social capital and knowledge contribution in electronic networks of practice. MIS Q. **29**, 35–57 (2005)
2. Wasko, M.M., Teigland, R., Faraj, S.: The provision of online public goods: examining social structure in an electronic network of practice. Decis. Support Syst. **47**, 254–265 (2009). doi:10.1016/j.dss.2009.02.012
3. Petty, R.E., Cacioppo, J.T.: The elaboration likelihood model of persuasion. Adv. Exp. Soc. Psychol. **19**, 123–205 (1986). doi:10.1016/S0065-2601(08)60214-2
4. Meservy, T.O., Jensen, M.L., Fadel, K.J.: Evaluation of competing candidate solutions in electronic networks of practice. Inf. Syst. Res. **25**, 15–34 (2014). doi:10.1287/isre.2013.0502
5. Fadel, K.J., Meservy, T.O., Jensen, M.L.: Exploring knowledge filtering processes in electronic networks of practice. J. Manag. Inf. Syst. **31**, 158–181 (2015). doi:10.1080/07421222.2014.1001262
6. Dimoka, A.: What does the brain tell us about trust and distrust? Evidence from a functional neuroimaging study. MIS Q. **34**, 373–396 (2010)
7. Goel, V., Buchel, C., Frith, C., Dolan, R.J.: Dissociation of mechanisms underlying syllogistic reasoning. Neuroimage **12**, 504–514 (2000). doi:10.1006/nimg.2000.0636
8. Goel, V., Dolan, R.J.: Explaining modulation of reasoning by belief. Cognition **87**, B11–B22 (2003). doi:10.1016/S0010-0277(02)00185-3
9. Tsujii, T., Watanabe, S.: Neural correlates of dual-task effect on belief-bias syllogistic reasoning: a near-infrared spectroscopy study. Brain Res. **1287**, 118–125 (2009). doi:10.1016/j.brainres.2009.06.080
10. Klucharev, V., Smidts, A., Fernández, G.: Brain mechanisms of persuasion: How "expert power" modulates memory and attitudes. Soc. Cogn. Affect. Neurosci. **3**, 353–366 (2008). doi:10.1093/scan/nsn022

11. Yarkoni, T.: NeuroSynth.org. doi:10.5281/zenodo.9925. http://neurosynth.org/locations/?y=22&x=52&z=12. Accessed 25 Apr 2016
12. Yarkoni, T.: NeuroSynth.org. doi:10.5281/zenodo.9925. http://neurosynth.org/locations/?y=34&x=4&z=-16. Accessed 25 Apr 2016
13. Cox, R.W.: AFNI: software for analysis and visualization of functional magnetic resonance neuroimages. Comput. Biomed. Res. **29**, 162–173 (1996)
14. Avants, B., Epstein, C., Grossman, M., Gee, J.: Symmetric diffeomorphic image registration with cross-correlation: evaluating automated labeling of elderly and neurodegenerative brain. Med. Image Anal. **12**, 26–41 (2008). doi:10.1016/j.media.2007.06.004
15. Klein, A., Andersson, J., Ardekani, B.A., et al.: Evaluation of 14 nonlinear deformation algorithms applied to human brain MRI registration. Neuroimage **46**, 786–802 (2009). doi:10.1016/j.neuroimage.2008.12.037
16. Forman, S.D., Cohen, J.D., Fitzgerald, M., et al.: Improved assessment of significant activation in functional magnetic resonance imaging (fMRI): use of a cluster-size threshold. Magn. Reson. Med. **33**, 636–647 (1995)
17. Xiong, J., Gao, J.H., Lancaster, J.L., Fox, P.T.: Clustered pixels analysis for functional MRI activation studies of the human brain. Hum. Brain Mapp. **3**, 287–301 (1995)
18. du Boisgueheneuc, F., Levy, R., Volle, E., et al.: Functions of the left superior frontal gyrus in humans: a lesion study. Brain **129**, 3315–3328 (2006). doi:10.1093/brain/awl244
19. Cutini, S., Scatturin, P., Menon, E., et al.: Selective activation of the superior frontal gyrus in task-switching: an event-related fNIRS study. Neuroimage **42**, 945–955 (2008). doi:10.1016/j.neuroimage.2008.05.013
20. Moss, H.E., Abdallah, S., Fletcher, P., et al.: Selecting among competing alternatives: selection and retrieval in the left inferior frontal gyrus. Cereb. Cortex **15**, 1723–1735 (2005). doi:10.1093/cercor/bhi049
21. Poldrack, R.A., Wagner, A.D., Prull, M.W., et al.: Functional specialization for semantic and phonological processing in the left inferior prefrontal cortex. Neuroimage **10**, 15–35 (1999). doi:10.1006/nimg.1999.0441
22. Costafreda, S.G., Fu, C.H.Y., Lee, L., et al.: A systematic review and quantitative appraisal of fMRI studies of verbal fluency: role of the left inferior frontal gyrus. Hum. Brain Mapp. **27**, 799–810 (2006). doi:10.1002/hbm.20221
23. Radua, J., Phillips, M.L., Russell, T., et al.: Neural response to specific components of fearful faces in healthy and schizophrenic adults. Neuroimage **49**, 939–946 (2010). doi:10.1016/j.neuroimage.2009.08.030
24. Caminiti, R., Ferraina, S., Johnson, P.B.: The sources of visual information to the primate frontal lobe: a novel role for the superior parietal lobule. Cereb. Cortex **6**, 319–328 (1996)
25. Vogt, B.A., Finch, D.M., Olson, C.R.: Functional heterogeneity in cingulate cortex: the anterior executive and posterior evaluative regions. Cereb. Cortex **2**, 435–443 (1992)
26. Bush, G., Vogt, B.A., Holmes, J., et al.: Dorsal anterior cingulate cortex: a role in reward-based decision making. Proc. Natl. Acad. Sci. U. S. A. **99**, 523–528 (2002). doi:10.1073/pnas.012470999
27. Bush, G., Luu, P., Posner, M.I.: Cognitive and emotional influences in anterior cingulate cortex. Trends Cogn. Sci. **4**, 215–222 (2000)

Helping an Old Workforce Interact with Modern IT: A NeuroIS Approach to Understanding Technostress and Technology Use in Older Workers

Stefan Tams and Kevin Hill

Abstract Older workers (defined here as 60 years of age and over) are experiencing major problems using modern information technologies (IT). These problems include greater anxiety and more stress when using IT. The proliferation of interruptions mediated by IT is especially problematic for them. Thus, this research has four objectives: (1) to develop a research model explaining which broad-spectrum cognitive mechanisms mediate the impact of age on stress in today's interruption age, (2) to explain the importance of having interruptions appear at predictable locations on the screen so as to help older workers use IT with greater ease and efficacy, (3) to explain the importance of using calm interruptions (i.e., no animation or aural alert) to help older workers be less stressed and more productive members of the organizations they work for, and (4) to offer practical, concrete guidance regarding interruption design and organization to software engineers and managers.

Keywords Interruptions • Technostress • Stress • Age • Older workers

1 Introduction

Anne Oldfellow approaches retirement age. To support her work as a sales manager, she has to use an ever-growing variety of ever more modern IT. Like many older workers, she has a positive attitude toward IT, but she faces trouble using it effectively. Anne is especially bothered by frequent interruptions such as unexpected email notifications, system notifications, and instant messages; she finds them disruptive to mental work, preventing her from remaining concentrated on her

S. Tams (✉)
Department of Information Technologies, HEC Montréal, Montréal, QC, Canada
e-mail: stefan.tams@hec.ca

K. Hill
Department of Human Resources Management, HEC Montréal, Montréal, QC, Canada
e-mail: kevin.hill@hec.ca

F.D. Davis et al. (eds.), *Information Systems and Neuroscience*, Lecture Notes in Information Systems and Organisation 16, DOI 10.1007/978-3-319-41402-7_3

task. Anne is frustrated and feels that IT threatens her well-being and job performance, assuming, "It's not for me, I am too old!" This vignette illustrates the interdependency of two emergent trends in organizations that warrant further study: the "graying" of the workforce and the pervasive growth of interruptions at work [1].

Today's workforce has to deal with countless interruptions mediated by IT, such as a constant stream of instant messages (through Skype for business), email notifications, meeting reminders, task reminders, text messages, etc. These technology-mediated (T-M) interruptions are coming from laptops, smartphones, and other devices, all of which beep, buzz, and blink relentlessly. Consequently, T-M interruptions constantly call for workers' attention. In fact, they have grown to consume about a third of employees' work days, causing stress [2–4]. The stress linked to IT, also termed technostress, can have hazardous consequences for employees in the forms of reduced psychological and physiological well-being (for example, negative emotions and elevated levels of stress hormones) [5–10]. As a result, workers' performance on IT-based tasks suffers [10].

At the same time, the workforce is aging rapidly; in fact, it is said to be *graying* [11–13]. According to the OECD, all of its current 34 member countries are experiencing rapid population and workforce aging [12–16]. The number of older workers has increased from 15,906,000 in the year 2000 to 24,548,000 in 2012 across the OECD countries, an increase of more than 50 % [17]. Importantly, the trend toward an older workforce will continue into the future; the participation of older people in the workforce is expected to increase sharply over the next several decades across the OECD member countries [11, 12].

As workers age, cognitive and biological changes occur. These changes include, for example, the reduced ability to ignore interruptions and remain concentrated on a task, challenges related to memory decline, and the reduced ability to operate computer input devices [18–20]. As a result of these changes, older workers are more at risk of being affected by technological stressors in general, and T-M interruptions in particular [1]. Yet, little guidance currently exists in the corresponding literatures for technology designers in terms of how to design IT with older workers in mind [21]. Such guidelines would improve older workers' well-being as well as their work performance and, thus, merit careful investigation.

This research will make theoretical and empirical contributions. *It will offer a causal model elucidating the process by which age-related impacts on technostress unfold, and it will test the model by means of moderated-mediation analysis.* The model will hypothesize that older workers experience more stress when using IT and encountering T-M interruptions than their younger counterparts due to the cognitive changes that occur with age, and it will help confirm that certain types of interruption presentation formats can moderate this problem for older workers. In doing so, this research will also contribute to the literature on interruption-based technostress by elucidating how and why such stress arises, by taking a design science approach to propose ways of moderating such stress, and by conceptualizing and measuring stress comprehensively using neuroscience methods and psychometrics. The combined use of neuroscience methods and psychometrics will help generate richer explanations for how interruption-based stress arises [10]. Finally, this research will advise practitioners on designing interfaces that older workers can use more effectively, a necessity for generations to come. With the dynamic nature of technological change that shows no sign of abating, older workers' disadvantages will be just as acute in the future as they are today. Overall, this research will confirm that one can lessen older workers' problems in using IT by manipulating the nature and characteristics of T-M interruptions, allowing older people to remain productive members of the workforce in today's interruption age. The next section provides the study background and develops our hypotheses. The third section outlines the general methodology to test our research model. The paper ends with concluding thoughts on its contributions.

2 Background and Hypotheses

Since this research will bring together different areas of inquiry, stress research and gerontology, it will draw on two complementary theoretical frameworks: person-environment fit (P-E fit) theory (i.e., a stress theory) and theories of cognitive aging.

P-E fit theory explains how the lack of fit between a person "P" and her environment "E" leads to stress [22, 23]. Both P and E are broadly conceptualized and can be used to represent numerous distinct concepts. In this study, the "P" represents a person's mental resource availability and the "E" represents such environmental demands as the complexity of the task environment. The degree of misfit between "P" and "E" is reflected in cognitive load, a construct that assesses the extent to which the capacity of a person's working memory is overloaded. Cognitive load can be measured physiologically using eye-tracking (tracks changes in eye movements as well as pupil dilation) and EEG technology (assesses changes in brain activity), or it can be measured psychologically using Likert-type scales (questions that respondents answer using predefined response options ranging from "strongly agree" to "strongly disagree") [24, 25]. Cognitive load increases with environmental demands or with cognitive decline, ultimately causing stress. It is particularly linked to stress in the case of excessive demands from, for example, difficulties in navigating complex interfaces in addition to the regular task demands [26]. P-E Fit theory allows us to hypothesize that the cognitive decline associated with increased age will intensify technostress by increasing the cognitive load associated with computer-based tasks. Further, it allows us to hypothesize that the strength of this indirect effect depends on the presentation format of T-M interruptions such that the effect will be stronger for interruptions that use attention-grabbing features. Specifically, the effect will be stronger for interruptions that appear at unpredictable locations on the screen in a non-calm manner (using animation and sound effects) than for interruptions that are calm (no animation or sound effects) and predictable in terms of where they will appear on the screen.

Theories of cognitive aging explain the changes in cognitive abilities that occur as people age. As such, these theories complement P-E fit theory that also focuses on cognition, and they are germane to this research since IT-based work is cognitively demanding [27]. Pertinent frameworks of cognitive aging for this research include the inhibitory theory of cognitive aging [28] and attentional capture effects across age groups [29, 30]. We will use these frameworks to explain why advanced age is, generally, only problematic for T-M interruptions that appear at unpredictable locations on the screen in a non-calm manner (using animation and sound), whereas it might be of no consequence for predictable, calm interruptions.

This research offers the following stress-related hypotheses (see Fig. 1; Table 1):

H1: *There is a positive correlation between age and stress.*

H2: *Cognitive load mediates the positive correlation between age and stress.*

H3: *Predictability of spatial location moderates the indirect effect of age on stress (via Cognitive load) such that this indirect effect will be weaker for interruptions that appear at predictable locations on the screen than for interruptions that appear at unpredictable locations.*

H4: *Interruption calmness moderates the indirect effect of age on stress (via Cognitive load) such that this indirect effect will be weaker for calm interruptions (*i.e.*, interruptions appearing without the use of animation or aural alerts) than for non-calm ones (appearing with animation and aural alert).*

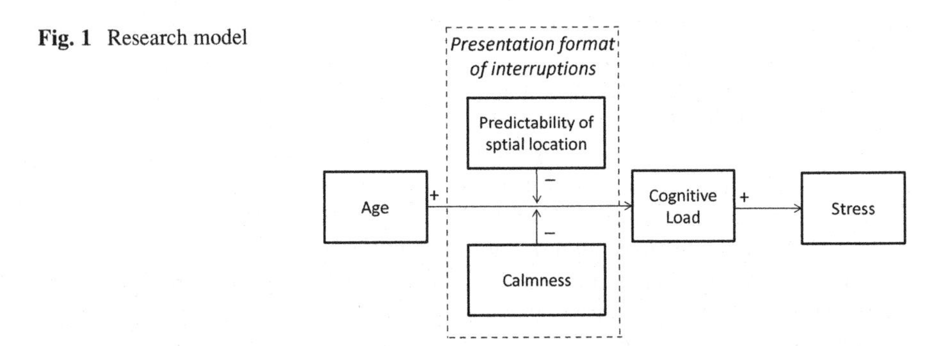

Fig. 1 Research model

Table 1 Construct definitions

Construct	Definition
Age	Chronologically younger compared to older individuals
Cognitive load	The mental load related to the executive control of working memory that increases with reduced cognitive abilities
Predictability of spatial location	Interruptions appearing at predictable locations on the screen versus interruptions appearing at unpredictable locations
Interruption calmness	Interruptions appearing with animation and sound effects versus interruptions appearing without such properties
Stress	The extent of psychological and physiological strain people experience as a result of their IT use

3 Methodology

To establish the link between age and cognitive load as well as stress in people who encounter interruptions with predetermined characteristics, an experiment will be conducted. A website will be created for the experiment on the basis of industry standards. The interrupting messages will, also, be designed on the basis of industry standards and validated by domain experts from practice to ensure ecological validity. Furthermore, manipulation checks will be performed to ensure that the interruptions reflect those that the participants commonly encounter in the real world. The experiment will employ a $2 \times 2 \times 2$ factorial design. The factors will be: age (younger [<30] vs. older [>60]), predictability of spatial location (predictable vs. unpredictable location on the screen), and interruption calmness (calm vs. animation and sound).

Unpredictable interruptions will reflect the common situation in which various types of interruptions appear at various locations all over the screen, while predictable ones will present a scenario in which various types of interruptions appear all at the same location on the screen, in the bottom right hand corner. Calm interruptions will appear on the screen in an unassuming way, without animation or aural alert. By contrast, non-calm interruptions will possess these attention-grabbing properties. Thus, for this initial study we plan on using an extreme manipulation of calmness that employs both animation and sound effects together so as to make the interruptions as intrusive as possible with the twin goals of maximizing treatment variance and increasing the efficiency of our experimental design (we will examine the individual effects of animation and sound in subsequent studies to tease apart how vision and hearing, which affect different brain regions and, thus, decline by different magnitudes as a function of age, might have differential impacts on the link between age and cognitive load [21]). We expect older adults to react significantly more strongly to our two moderating variables (i.e., predictability of spatial location and calmness).

Cognitive load will be assessed using oculometry and EEG technology. Regarding the former, a state of the art Tobii eye-tracker will be used to record changes in eye movements and pupil dilation, where increasing pupil dilation (rather than decreasing dilation) is generally indicative of increases in cognitive load. As to EEG technology, cognitive load can be inferred on the basis of a change in the spectral power of the EEG signals [31]. The theta-band power at the Fz electrode position has been shown to increase with increasing mental load, whereas the signals in the alpha-band have been found to decrease at Fz and Pz positions with increasing load [31]. Our experiment will employ an EGI wide-band EEG with 32-electrode headsets to record electrical activity along the scalp and permit us to analyze changes in the theta and alpha bands, ultimately indicating changes in cognitive load [24, 31].

To evaluate stress levels in our participants, salivary cortisol and α-amylase will be collected. These measures of stress are complementary in that they represent the two main systems responsible for the physiological aspects of stress: the autonomic nervous system (more specifically, the sympathetic-adrenomedullary system, whose main hormone is adrenaline) and the hypothalamic-pituitary-adrenal (HPA) axis [6, 10]. Cortisol is a stress hormone that represents the HPA axis and has to passively diffuse into saliva from the general circulation. By contrast, α-amylase is a digestive enzyme that is produced locally in the salivary glands. As such, it is a marker of the sympathetic nervous system component of the psychobiology of stress that reflects changes in the stress hormone adrenalin (produced by the sympathetic-adrenomedullary system). A-amylase can be found in saliva in relatively high concentrations. We will control for possible confounding factors, such as alcohol or caffeine consumption.

In addition, a behavioral performance measure similar to the one used by Tams et al. [10] will be employed. The resulting performance data will allow us to conduct a post-hoc analysis of how stress-related performance impacts differ between younger and older users, extending the results obtained by Tams et al. [10].

For data analysis, we will use Preacher et al.'s [32] approach to estimate the conditional indirect effects at different levels of the moderators. Specifically, Hypothesis 2 will be tested using Preacher and Hayes' [33] standard SPSS macro for testing indirect effects. Hypotheses 3 and 4 will be tested using Preacher et al.'s [32] standard SPSS macro for 1st stage moderated-mediation (i.e., Model 2 in [32]) to estimate the conditional indirect effects at different levels of the moderators.

4 Conclusion

The research proposed here will confirm that one can lessen older workers' problems in using IT by manipulating the characteristics of T-M interruptions. It will also suggest how training or corporate policies emphasizing the regulated use of interruptions may boost employee well-being and performance. The results expected from this research will inform the development of practical guidelines today and in the future. While the next generation of older workers will have grown up with today's IT and interruptions, the (new) technologies used then will, again, create problems for them. Our focus on cognitive aging permits us to develop guidelines rooted in cognition that will transcend specific generations of technologies and, thus, remain relevant.

Acknowledgments This research is being supported by the Fonds de recherche du Québec—Société et culture.

References

1. Tams, S.: The role of age in technology-induced workplace stress. Unpublished Doctoral Dissertation, Clemson University (2011)

2. Spira, J.B., Feintuch, J.B.: The Cost of Not Paying Attention: How Interruptions Impact Knowledge Worker Productivity. Basex, New York (2005)
3. Tams, S., Thatcher, J., Grover, V., Pak, R.: Selective attention as a protagonist in contemporary workplace stress: implications for the interruption age. Anxiety Stress Coping **28**(6), 663–686 (2015)
4. Tams, S., Thatcher, J., Ahuja, M.: The impact of interruptions on technology usage: exploring interdependencies between demands from interruptions, worker control, and role-based stress. In: Davis, F.D., et al. (eds.) Lecture Notes in Information Systems and Organisation: Information Systems and Neuroscience, vol. 10, pp. 19–26. Springer, Heidelberg (2015)
5. Ayyagari, R., Grover, V., Purvis, R.: Technostress: technological antecedents and implications. MIS Q. **35**(4), 831–858 (2011)
6. Riedl, R.: On the biology of technostress: literature review and research agenda. Database Adv. Inf. Syst. **44**(1), 18–55 (2013)
7. Riedl, R., Kindermann, H., Auinger, A., Javor, A.: Technostress from a neurobiological perspective—system breakdown increases the stress hormone cortisol in computer users. Bus. Inf. Syst. Eng. **4**(2), 61–69 (2012)
8. Riedl, R., Kindermann, H., Auinger, A., Javor, A.: Computer breakdown as a stress factor during task completion under time pressure: identifying gender differences based on skin conductance. Adv. Hum. Comput. Interac. **2013**, 1–8 (2013). doi:10.1155/2013/420169. Article ID 420169
9. Tams, S.: Challenges in technostress research: guiding future work. In: Proceedings of the 21st Americas Conference on Information Systems, Puerto Rico, Article 38, 7 pages (2015)
10. Tams, S., Hill, K., Ortiz de Guinea, A., Thatcher, J., Grover, V.: NeuroIS—alternative or complement to existing methods? Illustrating the holistic effects of neuroscience and self-reported data in the context of technostress research. J. Assoc. Inf. Syst. **15**(10), 723–753 (2014). Article 1
11. OECD: Work-force ageing in OECD countries. In: OECD Employment Outlook 1998, pp. 123–151. OECD, Paris (1998)
12. OECD: Helping older workers find and retain jobs. In: Pensions at a Glance 2011: Retirement-Income Systems in OECD and G20 Countries, pp. 67–79. OECD, Paris (2011)
13. Pham, S.: The Graying Work Force. The New York Times. Retrieved from http://newoldage.blogs.nytimes.com/2010/11/30/the-graying-workforce/?_php=true&_type=blogs&_r=0. Accessed 30 Nov 2010
14. OECD: The ageing challenge. In: Ageing and The Public Service: Human Resource Challenges, pp. 7–28. OECD, Paris (2007)
15. OECD: OECD Employment Outlook 2013. OECD, Paris (2013)
16. OECD: Editorial. In: International Migration Outlook 2013, pp. 9–10. OECD, Paris (2013)
17. OECD: Labour market statistics: labour force statistics by sex and age. In: OECD Employment and Labour Market Statistics (Database, 2010)
18. Charness, N., Holley, P., Feddon, J., Jastrzembski, T.: Light pen use and practice minimize age and hand performance different in pointing tasks. Hum. Factors **46**(3), 373–384 (2004)
19. Darowski, E.S., Helder, E., Zacks, R.T., Hasher, L., Hambrick, D.Z.: Age-related differences in cognition: the role of distraction control. Neuropsychology **22**(5), 638–644 (2008)
20. Salthouse, T.A., Babcock, R.L.: Decomposing adult age differences in working memory. Dev. Psychol. **27**(5), 763–776 (1991)
21. Tams, S., Grover, V., Thatcher, J.: Modern information technology in an old workforce: toward a strategic research agenda. J. Strateg. Inf. Syst. **23**(4), 284–304 (2014)
22. Edwards, J.R.: An examination of competing versions of the person-environment fit approach to stress. Acad. Manage. J. **39**(2), 292–339 (1996)
23. Siegrist, J.: Adverse health effects of high-effort/low-reward conditions. J. Occup. Health Psychol. **1**(1), 27–41 (1996)
24. Joseph, S.: Measuring cognitive load: a comparison of self-report and physiological methods. Unpublished Doctoral Dissertation, Arizona State University (2013)

25. Sweller, J.: Element interactivity and intrinsic, extraneous, and germane cognitive load. Educ. Psychol. Rev. **22**, 123–138 (2010)
26. Lazarus, R.S.: Stress and Emotion: A New Synthesis. Springer, New York (1999)
27. Gallivan, M.J., Spitler, V.K., Koufaris, M.: Does information technology training really matter? A social information processing analysis of coworkers' influence on IT usage in the workplace. J. Manag. Inf. Syst. **22**(1), 153–192 (2005)
28. Carlson, M.C., Hasher, L., Connelly, S.L., Zacks, R.T.: Aging, distraction, and the benefits of predictable location. Psychol. Aging **10**(3), 427–436 (1995)
29. Christ, S.E., Castel, A.D., Abrams, R.A.: Capture of attention by new motion in young and older adults. J. Gerontol. B Psychol. Sci. Soc. Sci. **63B**(2), P110–P116 (2008)
30. Pratt, J., Bellomo, C.N.: Attentional capture in younger and older adults. Aging Neuropsychol. Cognit. **6**(1), 19–31 (1999)
31. Müller-Putz, G.R., Riedl, R., Wriessnegger, S.C.: Electroencephalography (EEG) as a research tool in the information systems discipline: foundations, measurement, and applications. Commun. Assoc. Inf. Syst. **37** (2015). Article 46
32. Preacher, K.J., Rucker, D.D., Hayes, A.F.: Addressing moderated mediation hypotheses: theory, methods, and prescriptions. Multivar. Behav. Res. **42**(1), 185–227 (2007)
33. Preacher, K.J., Hayes, A.F.: SPSS and SAS procedures for estimating indirect effects in simple mediation models. Behav. Res. Methods Instrum. Comput. **36**(4), 717–731 (2004)

Mobile Multitasking Distraction: A Pilot Study with Intracranial Electroencephalography

Emma Campbell, Pierre-Majorique Léger, Élise Labonté-LeMoyne, Sylvain Sénécal, Marc Fredette, Franco Lepore, and Dang Nguyen

Abstract Texting while walking is a widespread and dangerous behaviour. Efforts are being put towards the development of mobile applications to refrain users from engaging in this behavior. The study of the neuropsychological mechanisms underlying this behavior will help developers aim specific cognitive processes. This study uses intracranial electroencephalography to increase our spatial understanding of the processes implicated in mobile multitasking. We asked a subject implanted with 128 electrodes to engage in a texting conversation while having to discriminate a point-light walker's direction. We discuss our results and methodological learning from this pilot study.

Keywords Intracranial electroencephalography • Electrocorticography • Mobile multitasking

---

E. Campbell (✉)
Department of Psychology, Centre de Recherche en Neuropsychologie et Cognition, University of Montreal, Montreal, QC, Canada

Tech3Lab, HEC Montreal, Montreal, QC, Canada
e-mail: emma.campbell@umontreal.ca

P.-M. Léger • É. Labonté-LeMoyne • S. Sénécal • M. Fredette
Tech3Lab, HEC Montreal, Montreal, QC, Canada
e-mail: pml@hec.ca; elise.labonte-lemoyne@hec.ca; ss@hec.ca; marc.fredette@hec.ca

F. Lepore
Department of Psychology, Centre de Recherche en Neuropsychologie et Cognition, University of Montreal, Montreal, QC, Canada
e-mail: franco.lepore@umontreal.ca

D. Nguyen
Centre Hospitalier Notre-Dame, Montreal, QC, Canada
e-mail: d.nguyen@umontreal.ca

F.D. Davis et al. (eds.), *Information Systems and Neuroscience*, Lecture Notes in Information Systems and Organisation 16, DOI 10.1007/978-3-319-41402-7_4

1 Introduction

Mobile multitasking (MMt) is defined as performing one or more information technology task while in movement. Texting while walking, a widespread and dangerous behaviour, is an instance of MMt. Injury reports related to MMt are becoming more important every year and are likely to keep increasing given the constantly growing use of smartphones [1]. While public safety research shows that mobile multitaskers are more cognitively distracted than non-mobile users, our team's recent work specifically measured this distraction using electroencephalography (EEG) [2]. While EEG results provide high temporal resolution on the phenomenon, we currently have limited spatial resolution. Greater spatial resolution would allow a better understanding of the neuropsychological mechanisms involved in this phenomenon.

In this paper, we present a pilot study that leverages the high temporal and spatial resolution of intracranial EEG, also known as electrocorticography (ECoG). To our knowledge is it the first time this method is applied in the field of NeuroIS [3]. The objective of this paper is to gain knowledge with regards to methodological challenges involved in using this method in an IS context. Also, the results will guide our investigation in identifying potential regions of interest for the next stage of this study. We conclude with a discussion on the advantages and limitations of ECoG for NeuroIS research.

2 Literature Review

2.1 Previous Work

Recently, the Tech^{3}Lab conducted what was, to our knowledge, the first EEG study on texting while walking [4]. It provided a better understanding of the neurocognitive processes that cause pedestrians to be distracted by texting and to place themselves in dangerous situations. In this study, participants were invited to walk on a treadmill and to discriminate the orientation of an oncoming human shape. In one condition, participants were simply walking, while in the other condition, they were walking and texting with a member of the research team. EEG was recorded in order to identify the neurocognitive processes that take place when participants have to disengage from texting to perform the discrimination task. It was found that participants performed worse on the discrimination task in the texting condition. It was also found that the correlation between an EEG index of engagement (beta/(alpha + theta) bands ratio) [5] and the discrimination task performance differed between conditions. A positive correlation was found when the participants were walking and texting and a negative one was observed when they were simply walking. These results suggest that MMt subjects seem more cognitively distracted than non-MMt subjects. While this study provided temporal resolution on the timing of the distraction, a method with a greater spatial resolution is needed to further understand the neuropsychological mechanisms involved.

2.2 *High Temporal and Spatial Resolution in Ecologically Valid MMt*

Advantages of ECoG Functional Magnetic Resonance Imaging (fMRI) has often been presented as a good alternative to overcome EEG's limitations in terms of spatial resolution. However, considering the static supine position required as well as the limitations on the presence of metal (i.e., no cellphones), adequate ecological validity would be impossible in this context. ECoG can provide, for our purposes, the required spatial and temporal resolution, while being ecologically valid. Although the basic principles are similar to scalp EEG, ECoG electrodes are connected directly to the subject's gray matter. Through the intervention of a neurosurgeon, the cranium is temporarily removed to implant a set of electrodes directly on the brain. ECoG is classically performed on patients suffering from a medication-resistant form of epilepsy. The intervention aims at precisely identifying the locus of the epileptic crisis for future surgery. After the implantation of the ECoG system, the patient is required to stay in the hospital for 2 weeks for continuous recording of the signal. In many teaching hospitals, neurosurgeons allow scientists to invite patients to take part in neurophysiological studies during their time in hospital care. Patients are under no obligation, but often feel privileged to provide a contribution to science and appreciate the distraction.

Limitations of ECoG The temporal resolution of ECoG combined with its spatial resolution is the golden standard of neurophysiological measuring. It can go as far as a single-neuron recording. Contrary to scalp EEG, the brain's electrical activity measured by ECoG originates from directly underneath the electrode. However, the medical setup necessary for such precision creates a limited environment for cognitive research [6]. Auditory and visual distractions are constantly present in the hospital setting. In view of the necessary set-up for clinical ECoG, it is not possible to move the subject to a quieter area. Another type of noise present in a hospital environment is electrical noise, thus the frequencies of interest must be chosen with care [7]. Finally, the electrodes placement is dependent on the patient's clinical needs and the spatial resolution of recordings of ECoG is limited to the position of the electrodes. Therefore, the brain regions of interest for the research question have to be in line with the location of the implanted electrodes.

These previous findings and methodological spatial limitations lead us to aim to better understand the cortical regions activated by task switching in a MMt context. MMt being a behaviour that can lead to undesirable outcomes and even dangerous consequences, it is crucial to identify the neural circuitry involved in this behavior.

This knowledge could assist the development of mobile applications aiming directly at the neural circuitry (e.g. stimulate the reward system if associated regions are implicated in MMt). This paper being, to our knowledge, the first ECoG project in NeuroIS, we primarily aim to observe the challenges of this method and list recommendations to this effect.

3 Methodology

3.1 Participant and Material

We conducted this pilot study with a participant recruited through collaborators at the Notre-Dame Hospital in Montreal, Canada who have an ethics certificate for cognitive studies. The participant was implanted with 128 electrodes in and around her temporal lobes (see Fig. 1). The data was acquired with a Stellate system (Stellate, Montreal, Canada) with no filter and an acquisition rate of 2000 Hz. The participant has extensive past experience with smartphones. The participant gave written consent to take part in the experiment. No compensation was given in exchange for her time.

3.2 Experimental Protocol

The stimulus was a point-light human shape performing a walking motion with an angle of 3.5° either to the right or to the left. The stimulus was presented on a computer screen (see Fig. 1). The subject was instructed to signal the walker's orientation (right or left) by saying her answer out loud. This task was used to

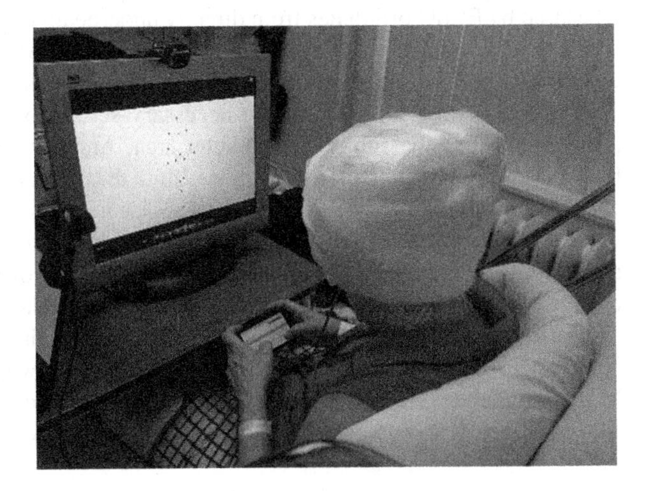

Fig. 1 Experimental set-up

represent a situation where an individual taking part in MMt would have to decide how to avoid an obstacle in an ecological setting. The appearances of the stimulus were announced by an audio cue at which point the subject had to make a decision about the orientation of the walker. Each task lasted 22 min including 44 walkers per task for a grand total of 88 trials.

Building upon Courtemanche et al. [4], the experimental session was divided into two conditions. In the first condition, the participant was asked to use the provided smartphone (iPhone 4) to exchange text messages with the research assistant. Conversation topics were predetermined, but flexible to let the conversation take its natural course. When she heard an audio cue, she was asked to determine the orientation of the oncoming stimulus. In the second condition, she was asked to simply look in front of her and wait for the stimulus to appear. A 5 min practice run for the first task was performed to demonstrate the instructions. A break between the two tasks was offered to the participant.

4 Results

The preliminary analyses were done using BrainVision Analyzer (Brain Products, Munich, Germany). ECoG channels with epileptic artefacts were removed. Epileptic artefacts are high frequency (80–500 Hz) spontaneous activity. They are not manifestations of an epileptic seizure, but are often found near the seizure onset zone.

The 63 channels that did not show epileptic artefacts were re-referenced to the common average reference and filtered with a 0.5–70 Hz bandpass filter with a 24 dB/oct slope. A 60 Hz notch filter was also applied. The data was downsampled to 1024 Hz, segmented into 1 s epochs and an FFT was applied with a resolution of 1 Hz. All epochs from a condition were averaged together. The engagement ratio of these averages was calculated with the power of each frequency band: $\beta/(\alpha+\theta)$ with beta 13–20 Hz, alpha 8–12 Hz and theta 4–7 Hz. The difference between the tasks (texting–control) was calculated for each electrode individually. Ongoing analyses are addressing the time-frequency patterns of the electrodes in order to observe frequency changes throughout the switches between the texting task and the discrimination task.

Preliminary results show that the engagement ratio was greater in the texting condition for almost all of the electrodes. The top 20 % of electrodes presenting the highest difference in the engagement ratio between the two tasks, being greater in the texting condition, were identified and plotted as can be seen in Fig. 2.

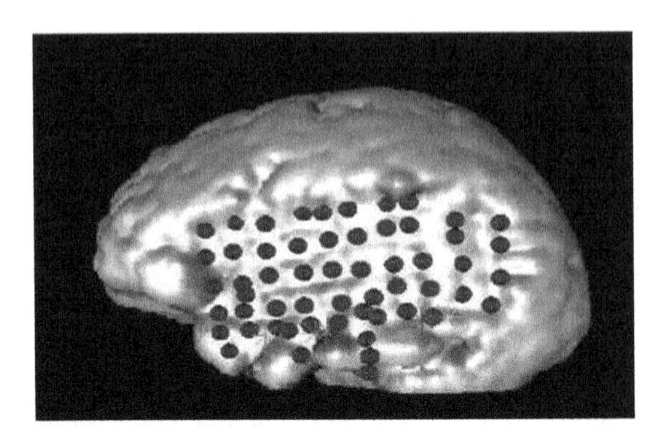

Fig. 2 Activated electrodes in texting condition

5 Discussion and Conclusion

Results show that activation increases in language processing areas during the texting condition. These results are consistent with the texting behavior. However, only one subject is not enough to develop a rich understanding of the phenomena and these results should mostly be used for illustrative purposes of the methodology. Indeed, this pilot study allowed the research team to acquire important methodological knowledge in using ECoG in an IS context. Our next step is to collect several other subjects over the next months.

Even though it comes with its load of challenges, ECoG remains the most precise neuroimaging method, both spatially and temporally. This task was performed to better our understanding of the task switching cost in a texting vs. discrimination task. Our understanding of task switching in a mobile context will involve the EEG and behavioural results and those acquired in previous and future work. In the specific context of texting while walking, where there is a health hazard to the users, it is imperative to understand which neural circuitry is implicated as it will influence the development of future applications and measures to counter the negative effects of this dangerous habit.

References

1. Léger, P.-M., Sénécal, S., Cameron, A.-F., Bellavance, F., Faubert, J., Fredette, M.: Travailler à l'extérieur des frontières de l'organisation: Un modèle pour étudier les effets des multitâches technologiques en contexte piétonnier. In: Proceedings of the Congrès de l'Association des Sciences Administratives du Canada (ASAC), Alberta (2013)
2. Courtemanche, F., Léger, P.-L., Cameron, A.-F., Faubert, J., Labonté-Lemoyne, É., Sénécal, S., Fredette, M., Bellavance, S.: Texting while walking: measuring the impact on pedestrian visual attention. In: Gmunden Retreat on NeuroIS 2014, Austria (2014)
3. Riedl, R., Léger, P.-M.: Tools in NeuroIS research: an overview. In: Fundamentals of NeuroIS, pp. 47–72. Springer, Heidelberg (2016)

4. Courtemanche, F., Labonté-Lemoyne, E., Léger, P.-M., Cameron, A.-F., Fredette, M., Faubert, J., Sénécal, S., Bellavance, F.: Texting while walking: an expensive switching cost. PLoS ONE (resubmitted)
5. Pope, A.T., Bogart, E.H., Bartolome, D.S.: Biocybernetic system evaluates indices of operator engagement in automated task. Biol. Psychol. **40**(1–2), 187–195 (1995)
6. Lachaux, J.P., Rudrauf, D., Kahane, P.: Intracranial EEG and human brain mapping. J. Physiol. Paris **97**(4–6), 613–628 (2003)
7. Lachaux, J.-P., Axmacher, N., Mormann, F., Halgren, E., Crone, N.E.: High-frequency neural activity and human cognition: past, present and possible future of intracranial EEG research. Prog. Neurobiol. **98**(3), 279–301 (2012)

Impact of Cognitive Workload and Emotional Arousal on Performance in Cooperative and Competitive Interactions

Anuja Hariharan, Verena Dorner, and Marc T.P. Adam

Abstract We examine whether changes in the social environment (competitive or cooperative), affect the relationship between the internal state (cognitive workload and emotional arousal), and the performance in a given task. In a controlled experimental game setting, participants played cooperatively and competitively with different partners. EEG activity and heart rate changes measured cognitive workload and emotional arousal respectively. Cognitive workload was associated negatively with performance in the competitive but not the cooperative mode. By contrast, arousal was associated negatively with performance in the cooperative mode but not the competitive mode.

Keywords Competition • Cooperation • EEG • Heart rate • NeuroIS

1 Introduction

Encouraging people to contribute to online communities, such as wikis and crowd working platforms, requires creating an engaging environment [1]. Encouraging them to contribute high-quality content requires an understanding of how the environment affects people's performance and how environment, performance and people's internal states are interrelated [2]. Performance in groups depends, among other environmental factors, on the mode of social interaction [3, 4]. For instance, [5] observed that elements of competition and collaboration helped to sustain the crowd's motivation to participate, and to produce quality design outcomes. [6] have shown that team formation is an important determinant of player engagement, and assessment of ideas, while employing serious games. By employing clustering methods, [7] showed methods of characterizing group player profiles, which may be used to achieve specific objectives, such as engaging players

A. Hariharan (✉) • V. Dorner
Karlsruhe Institute of Technology, Karlsruhe, Germany
e-mail: anuja.hariharan@kit.edu; verena.dorner@kit.edu

M.T.P. Adam
The University of Newcastle, Newcastle, Australia
e-mail: marc.adam@newcastle.edu.au

F.D. Davis et al. (eds.), *Information Systems and Neuroscience*, Lecture Notes in Information Systems and Organisation 16, DOI 10.1007/978-3-319-41402-7_5

with the serious goals of the game, when they do not engage equally as the others. While it has been highlighted that team formation matters for engagement and performance, it has not been researched how the social interaction mode (i.e., whether cooperative or competitive) is related to the neurophysiological aspects of performance. This work hence explores the relationship between behavior (game performance) and neurophysiological processes (emotional arousal and cognitive workload), in the context of a game that was played in cooperative and competitive interaction mode. Emotional arousal and cognitive workload were measured with heart rate changes and EEG brain activity as proxies.

2 Method

2.1 Procedure

We examined behavior in a pattern recognition game which participants played cooperatively (i.e., contributed to a common pool and shared payoff), and competitively (i.e., one's payoff harms the other). Every session comprised four phases including the initial rest phase (1). Then participants played in the individual mode to familiarize themselves with the game (2). In treatment phases (3) and (4), participants were matched with different partners in a perfect stranger matching to play in cooperative and competitive modes. Each half of the participants were assigned to different treatment orders. Each treatment consisted of 42 pattern matching tasks, and was played for 5 min ending with a result screen informing the participants of their game score and payoffs. A pause of 30 s between treatments gave participants the opportunity to mentally disengage from the preceding treatment and prepare for the next. After the second treatment, participants were shown another screen with their overall payoff from both treatments. Participants were paid out based on their game earnings (computed as 5 % of the cumulative earnings in both treatments).

During the entire experiment, brain activity of participants was recorded using a 32 channel EEG device (Actichamp, Version 2, by Brain Products), and heart rate data was recorded using Bioplux [8]) system. The study was conducted on 156 participants (76 male and 80 female, Bachelor or Master Students, without restriction of major). Participants were recruited using ORSEE [9], filtering on gender, and were invited to one of all-male or all-female sessions. The study was conducted in compliance with the university's ethics guidelines.

2.2 Measures

For the physiological measurement of cognitive workload, we adopted a standard procedure used in the context of measuring engagement, workload and attention with EEG data [10–12]. EEG data recorded during game play was used to compute the workload index. The procedure by [13], and applied in [14] was relied on: EEG data was sampled at 500 Hz with a band pass filter from 1 Hz to 40 Hz. The computation of cognitive workload index was executed in two steps: (1) computation of independent components and (2) computation of spectral powers in the 1–40 Hz frequency bands. Independent components were computed using fast Independent Component Analysis (fastICA) algorithm available in the EEGLab toolbox for 14 frontal channels for each subject, resulting in 14 independent components, which were artefact cleaned for each subject by eyeball inspection. Next, power spectra were calculated for 2 second windows before each event, across each treatment type for the event of clicking of a matching pattern by the subject. The EEG cognitive workload index mirrors the theoretical definition of cognitive workload, taking into account the absolute and relative power spectra from 1 to 30 Hz of EEG channel. The spectral powers of each of the frequency bands (Beta, Alpha, and Theta) are calculated on the independent components obtained from 14 frontal channels, and then cognitive workload is computed by the formula (Beta/(Alpha + Theta)) spectral power [13].

Emotional arousal (θHR) during the game is assessed by calculating the participant's average heart rate over the course of five seconds directly prior to a pattern-matching click in the game (cf. [15]). To rule out subject-specific effects, this value was normalized using the subject's base heart rate, as measured during the initial cool down period before the actual experiment. Table 1 summarizes the statistics for the observed values of emotional arousal and cognitive workload (cf. [16]).

3 Results

We first investigated how collaborative and competitive structures affect the relationship between cognitive workload and game performance. Game performance was measured by the variable "hit", which indicates the number of correct pattern

Table 1 Summary of observed values for cognitive workload and emotional arousal

Factor	Measurement statistics Cooperative mode	Measurement statistics Competitive mode
Emotional arousal (θHR)	M = 1.027 [SD = 0.146], min = 0.004, max = 2.373, median = 1.020	M = 0.984 [SD = 0.201], min = 0.004, max = 2.375, median = 1.001
Cognitive workload (CW Index)	M = 0.589 [SD = 0.295], min = 0.104, max = 4.386, median = 0.541	M = 0.571 [SD = 0.228], min = 0.113, max = 3.277, median = 0.543

matches by the given player, in either mode. Our data show that the expended level of cognitive workload was about the same across the experiment, indicating that all pattern matching tasks were of similar difficulty. A second check for differences in difficulty correlating the level of cognitive workload with the target position of patterns on the screen, also showed no differences. The experienced cognitive workload differences are hence likely to be due to the social interaction mode. A paired sample *t*-test revealed that differences in experienced cognitive workload were significant across treatments ($t\,(3177) = 1.908, p < .05$), with higher workload for cooperative mode than competitive mode.

Table 2 presents the generalized linear model results of performance in cooperative and competitive modes, as well as considering cooperative and competitive mode together. Here, "CW Index", represents the experienced cognitive workload index as computed by EEG. The cognitive workload index was negatively correlated with game performance in competitive but not in cooperative mode (columns 2 and 3 in Table 2). Taking both modes together and adding a dummy variable for game mode (Coop = 1, Comp = 0) into the regression show that the above results are robust. While a high EEG index corresponded with lower game performance in both modes, the interaction variables indicates cognitive workload lowered performance only in the competitive but not in the cooperative mode (column 4 in Table 2).

Second, we looked at whether the social mode affects the relation between experienced arousal (normalized average heart rate 5 s prior to click) and game performance. A paired sample t-test revealed that differences in experienced arousal were significant across treatments ($t\ (5297) = -8.875,\ p < .001$), with higher arousal for competitive mode. A generalized linear model regression is presented in Table 3. In cooperative mode, higher heart rate was associated with lower game performance. In competitive mode, heart rate had an insignificant positive relation to game performance. "Round" in all cases was significantly negatively related to game performance, albeit with low coefficients. This means that, over time, participants had fewer hits in both cooperative and competitive modes. In addition, over the course of the game, heart rates decreased very slightly, possibly because the games were played sitting down and participants' bodies

Table 2 Game mode-wise regressions for game performance (measured as hits)

	Hit (Coop)	Hit (Comp)	Hit (Coop & Comp)
Target pattern position #1–20	−0.007 (0.418)	−0.010 (0.280)	−0.009 (0.180)
Round (# 1–42)	0.003 (0.947)	0.005 (0.389)	0.002 (0.509)
CW Index (0–1.0)	0.127 (0.510)	−0.757 (0.001)**	−0.758 (0.001)**
Mode Coop (Dummy:0/1)			−0.351 (0.067)+
Mode Coop × CW Index			0.883 (0.004)**
Constant	0.754 (0.000)***	1.056 (0.000)***	1.081 (0.000)***
AIC	2009.1	2042	4047.4
N	1601	1577	3178

Regression coefficients, with standard errors in parentheses
+$p < 0.10$; *$p < 0.05$; **$p < 0.01$; ***$p < 0.001$

Table 3 Game mode-wise regressions for game performance (measured as hits)

	Hit (Coop)	Hit (Comp)	Hit (Coop &Comp)
Target pattern position #1–20	−0.006 (0.443)	−0.014 (0.062) +	−0.010 (0.063)+
Round #1–42	−0.015(0.0003)***	−0.009 (0.024)*	−0.012 (0.000)***
θHR (ratio, 0.5–2.0)	−0.955(0.0002)***	0.283 (0.345)	0.299 (0.317)
Mode Coop (Dummy:0/1)			1.260 (0.002)**
θHR × Mode Coop			−1.228 (0.002)**
Constant	2.390 (0.000)***	1.084(0.001)***	1.085 (0.001)***
AIC	2961.1	3031.9	5990.6
N	2644	2654	5298

Regression coefficients, with standard errors in parentheses
+$p < 0.10$; *$p < 0.05$; **$p < 0.01$; ***$p < 0.001$

Table 4 Game mode-wise regressions for game performance (measured as unclicked)

	Unclicked (Coop)	Unclicked (Comp)	Unclicked (Coop & Comp)
Target pattern position #1–20	0.017 (0.010)+	0.009 (0.009)	0.013 (0.007)+
Round #1–42	−0.002 (0.006)	−0.005 (0.006)	−0.004 (0.004)
CW Index (0–1.0)	−0.053 (0.208)	0.614 (0.227)**	0.615 (0.227)**
Mode Coop (Dummy:0/1)			0.203 (0.198)
Mode Coop × CW Index			−0.670 (0.309)*
Constant	−1.295 (0.197) ***	−1.360 (0.198) ***	−1.427 (0.174)***
AIC	1740.5	1826.8	3563.8
N	1601	1577	3178

Regression coefficients, with standard errors in parentheses
+$p < 0.10$; *$p < 0.05$; **$p < 0.01$; ***$p < 0.001$

needed less energy (c.f. [17]; heart rates ceteris paribus always go down during the experiment). Such a declining trend was not visible in the cognitive workload data. The third regression investigated the differences in emotional arousal on game performance between cooperative mode and competitive mode. Higher arousal in cooperative mode was associated with lower performance than competitive mode (column 4 in Table 3).

Finally, we investigated whether the tendency not to click on any pattern in a particular round is associated with cognitive workload and emotional arousal, depending on social interaction mode (Tables 3 and 4). In competitive mode, higher cognitive workload corresponded to a significantly higher number of unclicked decisions ($B = .614$, $SE = .227$, $p < .01$). This was also reflected when considering the data together, and controlling for interaction effects of mode and cognitive workload (Table 4).

The effect of arousal differs between treatments: in cooperative mode, higher arousal was associated with more unclicked, but in competitive mode, with fewer

Table 5 Game mode-wise regressions for game performance (measured as unclicked)

	Unclicked (Coop)	Unclicked (Comp)	Unclicked (Coop & Comp)
Target pattern position (#1–20)	0.016 (0.009)+	0.019 (0.009)*	0.018 (0.006)**
Round #1–42	0.019 (0.005)***	0.004 (0.005)	0.011 (0.003)**
Arousal	1.040 (0.315)***	−0.713 (0.343)*	−0.748 (0.342)*
Mode Coop (Dummy:0/1)			−1.742 (0.476)***
Mode Coop × Index			1.721 (0.465)***
Constant	−3.272 (0.362) ***	−1.219 (0.373) **	−1.318 (0.363)***
AIC	2262.2	2313.6	4576.7
N	1601	1577	3178

Regression coefficients, with standard errors in parentheses
$+p < 0.10$; $*p < 0.05$; $**p < 0.01$; $***p < 0.001$

unclicked (Table 5). Thus, in competitive mode, arousal was associated with higher engagement, whereas in cooperative mode, it was associated with lower engagement.

4 Summary

We examined the impact of cognitive workload and emotional arousal on competitive performance, in comparison to a cooperative mode. An experiment was conducted where participants played a pattern recognition game cooperatively and competitively, and physiological measures of EEG and heart rate were measured. Our results indicate that in competitive mode, higher levels of cognitive workload are associated with lower performance. These results suggest that the experience of high cognitive workload in competitive social mode might not be necessarily beneficial, or in other words, lesser deliberation might lead to an improved performance. These results are complementary to those of [18], which suggest that cognitive absorption was positively related to a more relaxed, less vigilant state of the learner. In the next steps, in addition to linear effects, quadratic effects of cognitive workload on participant behavior would need to be compared as well. In the cooperative mode, a higher arousal level was associated with poorer performance, or viewed otherwise, people who do not perform well are potentially experiencing an increase in heart rate due to anger, or frustration experienced. In contrast, the participants who perform well are possibly in a state of "flow" [19, 20] and are experiencing an optimal level of arousal.

Finally, the tendency to not click a pattern in a particular round, in this gaming context was explained by cognitive workload (increasing it in competitive mode) and emotional arousal (increasing/decreasing it in cooperative/competitive mode respectively). These results suggest that in competitive mode, cognitive workload can manifest as lack of participation in a given setting. Emotional arousal has the opposite effect on propensity to click. Since this study has been conducted in a gaming content, not subject to a specific information framing, these will need to be extrapolated to specific social environments, such as Wikipedia contributions (as a specific case of collaborative), and crowd work (to address whether to design a task cooperatively or competitively). The willingness to expend effort in such cooperative and competitive modes is also relevant for understanding participatory behavior and willingness to contribute in an online social context.

References

1. Preece, J., Shneiderman, B.: The reader-to-leader framework: motivating technology-mediated social participation. AIS Trans Hum Comp Inter **1**(1), 13–32 (2009)
2. Galegher, J., Kraut, R.E., Egido, C.: Intellectual Teamwork: Social and Technological Foundations of Cooperative Work. Psychology Press (2014)
3. Malhotra, D.: The desire to win: the effects of competitive arousal on motivation and behavior. Organ Behav Hum Decis Process **111**(2), 139–146 (2010)
4. Adam, M.T.P., Kroll, E.B., Teubner, T.: A note on coupled lotteries. Econ Lett **124**(1), 96–99 (2014)
5. Park, C.H., Son, K., Lee, J.H., Bae, S.H. Crowd vs. crowd: large-scale cooperative design through open team competition. Proceedings of the 2013 Conference on Computer Supported Cooperative Work, 1275–1284 (2013)
6. Feldmann, N., Adam, M.T.P., Bauer, M. Using serious games for idea assessment in service innovation: team formation matters. ECIS 2014 Proceedings, Tel Aviv, Israel, pp. 1–17 (2014)
7. Cornforth, D., Adam, M.T.P. Clustering evaluation, description, and interpretation for serious games. In: Loh, C.S., Sheng, Y., Ifenthaler, D. (eds.) Serious Games Analytics, pp. 135–155. Springer. ISBN 978-3-319-05833-7 (2015)
8. Bioplux PLUX – Wireless Biosignals. [online] http://www.plux.info/systems (2007). Accessed 14 July 2011
9. Greiner, B. The online recruitment system ORSEE 2.0: a guide for the organization of experiments in economics. In: Kremer, K., Macho, V. (eds.) Forschung und wissenschaftliches Rechnen, 2003, pp. 79–93. GWDG Bericht 63 [Research and Scientific Computing] (2004)
10. Hariharan, A., Adam, M.T.P., Teubner, T., Weinhardt, C.: Think, feel, bid: the impact of environmental conditions on the role of bidders' cognitive and affective processes in auction bidding. Electron. Markets, 1–17 (2016)
11. Ortiz de Guinea, A., Titah, R., Léger, P.-M.: Measure for measure: a two study multi-trait multi-method investigation of construct validity in IS research. Comp Hum Behav **29**(3), 833–844 (2013)
12. Ortiz de Guinea, A., Titah, R., Léger, P.-M.: Explicit and implicit antecedents of users' behavioral beliefs in information systems: a neuropsychological investigation. J Manage Inf Syst **30**(4), 179–210 (2014)
13. Pope, A.T., Bogart, E.H., Bartolome, D.S.: Biocybernetic system evaluates indices of operator engagement in automated task. Biol Psychol **40**(1), 187–195 (1995)
14. Charland, P., Léger, P.-M., Sénécal, S., Courtemanche, F., Mercier, J., Skelling, Y., Labonté-Lemoyne, E.: Assessing the multiple dimensions of engagement to characterize learning: a neurophysiological perspective. J Vis Exp **101**, e52627 (2015)

15. Teubner, T., Adam, M.T.P., Riordan, R.: The impact of computerized agents on immediate emotions, overall arousal and bidding behavior in electronic auctions. J Assoc Inf Syst **16**(10), 838–879 (2015)
16. Charland, P., Allaire-Duquette, G., Léger, P.-M.: Collecting neurophysiological data to investigate users' cognitive states during game play. J Comput **2**(3), 20–24 (2014)
17. Adam, M.T.P., Krämer, J., Müller, M.B.: Auction fever! How time pressure and social competition affect bidders' arousal and bids in retail auctions. J Retail **91**(3), 468–485 (2015)
18. Léger, P.M., Davis, F.D., Cronan, T.P., Perret, J.: Neurophysiological correlates of cognitive absorption in an enactive training context. Comput Hum Behav **34**, 273–283 (2014)
19. Csikszentmihalyi, M.: Flow: the psychology of optimal experience, vol. 41. Harper Perennial, New York, NY (1991)
20. Bastarache-Roberge, M.-C., Léger, P.-M., Courtemanche, F., Sénécal, S., Fredette, M. Measuring flow using psychophysiological data in a multiplayer gaming context. In: Information Systems and Neuroscience, pp. 187–191. Springer (2015)

It All Blurs Together: How the Effects of Habituation Generalize Across System Notifications and Security Warnings

Bonnie Brinton Anderson, Anthony Vance, Jeffrey L. Jenkins, C. Brock Kirwan, and Daniel Bjornn

Abstract Habituation to security warnings—the diminished response to a warning with repeated exposures—is a well-recognized problem in information security. However, the scope of this problem may actually be much greater than previously thought because of the neurobiological phenomenon of generalization. Whereas habituation describes a diminished response with repetitions of the same stimulus, *generalization* occurs when habituation to one stimulus carries over to other novel stimuli that are similar in appearance. Because software user interface guidelines call for visual consistency, many notifications and warnings share a similar appearance. Unfortunately, generalization suggests that users may already be deeply habituated to a warning they have never seen before because of exposure to other notifications. In this work-in-progress study, we propose an eye tracking and fMRI experiment to examine how habituation to frequent software notifications generalizes to infrequent security warnings, and how security warnings can be designed to resist the effects of generalization.

Keywords Security warnings • Habituation • Generalization • fMRI • Eye tracking • NeuroIS

1 Introduction

Habituation is a major factor in users' failure to adhere to security warnings [1, 2]. Through this neurological phenomenon, repeated warnings that once were salient become virtually unnoticeable. *Generalization* occurs when the effects of habituation to a repeated stimulus carry over to other novel stimuli that are similar in appearance [3]. Applied to the domain of information security, generalization suggests that users not only habituate to individual security warnings, but also to whole classes of notifications and warnings that share a similar appearance and user

B.B. Anderson (✉) • A. Vance • J.L. Jenkins • C.B. Kirwan • D. Bjornn
Brigham Young University, Provo, UT, USA
e-mail: bonnie_anderson@byu.edu; anthony.vance@byu.edu; jeffrey_jenkins@byu.edu; kirwan@byu.edu; daniel.bjornn@gmail.com

F.D. Davis et al. (eds.), *Information Systems and Neuroscience*, Lecture Notes in Information Systems and Organisation 16, DOI 10.1007/978-3-319-41402-7_6

interaction (UI) paradigm (see Fig. 1). If true, then the threat and potential impact of habituation is much broader than previous work has suggested [4–6], as users may already be deeply habituated to a security warning that they have never seen before.

We outline an experiment using eye tracking and fMRI to (1) measure the extent to which the effects of habituation generalize across similar types of notifications and security warnings, and (2) determine warning designs that can reduce the occurrence of generalization.

2 Literature Review

2.1 Habituation and Generalization

Habituation is defined as the decreased response to a repeated stimulation [7]. Habituation is fundamentally a neurobiological phenomenon [8], and is evident "in every organism studied, from single-celled protozoa, to insects, fish, rats, and people" ([9], p. 125). Not surprisingly, humans also exhibit habituation to a wide variety of stimuli as early as infancy [10].

Habituation with novel stimuli can have one of two responses: stimulus specificity, or stimulus generalization. With stimulus specificity, a novel stimulus similar to the habituated stimulus, will not show a habituated response [11]. However, the fields of neuroscience and neurobiology show that the effects of habituation to a familiar stimulus can generalize to other, novel (i.e., never before experienced) stimuli that share similar characteristics [3, 12]. This occurs for animals with simple stimuli [13] as well as for people with complex, real-world stimuli such as video games [11]. This generalization means that although the stimulus is novel, the response is similar to a habituated response to a repeated stimulus. When habituation occurs, generalization is also likely to occur.

2.2 Habituation and Generalization to Security Warnings

Given its strong security implications, habituation is frequently cited as a key contributor to users' failure to heed warnings. However, many studies infer the

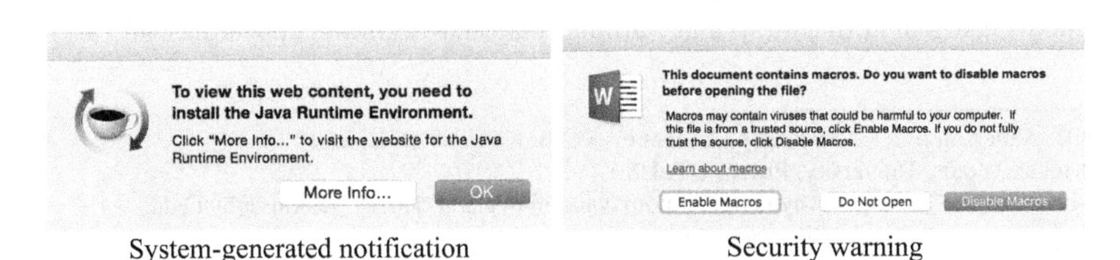

System-generated notification Security warning

Fig. 1 A notification and security warning. Note the similarities in UI and mode of interaction

presence of habituation, rather than empirically examine it. For example, Sunshine et al. [14] found a correlation between user disregard for warnings and user recognition of warnings as previously viewed, and attributed this correlation to habituation. An exception is Anderson et al. [4], who used fMRI to measure habituation in the brain in response to warnings. Their results showed a large drop in activity in the visual processing centers of the brain after only the second exposure to a warning, and found further decreases with additional exposures.

Generalization of habituation effects has important implications for security warnings. Although users are regularly presented with a variety of warnings [5], research on human–computer interaction indicates that the frequency of warnings is small compared with the barrage of system-generated notifications that are common to the computing experience [15, 16]. The high frequency of notifications is compounded by UI guidelines that call for visual consistency across UI elements such as notifications [12]. If habituation to these notifications generalize, users are unlikely to give full attention to security warnings even if they haven't seen them before [9]. This problem can be exacerbated by the common software engineering practice of consistency across messages. In the context of security messages, we believe that clarity should trump consistency.

The effects of generalization of software notifications to warnings has not been empirically examined previously. Instead, researchers have observed that habituation to a single warning in one context carries over to a different context. For example, Sunshine et al. [14] observed that users who correctly identified the risks of an SSL warning in a library context inappropriately identified these same risks in a banking context. Similarly, Amer et al. [17] found that users who habituated to exception notifications in one context were habituated to a different through visually identical exception notification in a different context. However, in each of these cases, users habituated to the same type of security warning or notification. As a result, it is unclear to what extent software notifications generalize to security warnings.

2.3 Hypotheses

Replicating the findings of previous studies [18], we hypothesize that users will habituate to software notifications. When users repeatedly see software notifications, the brain creates a mental model of these notifications. Rather than giving attention to future exposures to the notifications, the brain increasingly relies on this mental model. As a result, users' responses to future notifications decrease (i.e., habituate) in response to repeated exposures of notifications [19]. In summary, we predict:

H1 Users will habituate to repeated exposures of software notifications.

We predict that habituation to software notifications will also decrease users' responses to security warnings. Habituation to stimuli extends beyond the specific sensory characteristics of a repeated stimulus. Rather, habituation is also "happening centrally rather than in primary sensory afferents" ([3], p. 137). In other words, when a person sees and habituates to repeated software notifications, this decreases one's response to other stimuli in the environment. People's responses to security warnings may be particularly vulnerable to the effects of software notification habituation because they share several of the same properties (e.g., both popup on a computer screen, they often share the same shape or size, and some even have the same look and feel). As such, the brain may rely on the mental models of software notifications when interpreting security messages, which will result in generalized habituation to the security messages [19]. In summary, we predict:

H2 Habituation to software notifications will generalize to security warnings.

To test for generalization, we will present several stimuli equally capable of eliciting a response. We will compare whether the response to the first novel stimulus is similar to the first exposure original stimulus. If the novel stimulus response is weaker than the first exposure of the original stimulus, then there is evidence of generalization. However, if the novel stimulus response is greater than the first exposure of the original stimulus, that provides evidence of stimulus specificity habituation [11].

3 Experimental Design

3.1 Methodology

We plan to use fMRI and eye-tracking tools simultaneously. We are interested in capturing data that will allow us to examine the levels of cognitive processing, thus requiring a neurophysiological tool such as fMRI that has high spatial resolution of brain activity. While fMRI has the highest spatial resolution of any non-invasive volumetric brain imaging technology [20], tools such as eye-tracking have a much higher temporal resolution. In addition, by using these two tools concurrently, we can examine the eye fixations and pair them with the neural responses. We will use the Siemens 3 T Tim-Trio scanner at our university's MRI research facility. In addition, we will collect eye-tracking data during each scan using an MRI-compatible SR Research EyeLink 1000 Plus long-range eye tracker. Further, using eye-tracking technology will assist future research seeking to improve the design elements of security warnings by specifying which areas of interest are critical.

3.2 Task

The task will use a common go/no-go design [21], in which participants will be instructed to respond to items they see both as quickly and as accurately as possible by pressing a button on a button box. Go trials, or trials in which participants must press a button each time the stimulus is presented, will consist of images of notifications. No-go trials, or trials in which participants must inhibit the tendency to press a button and withhold their response, will be made up of warnings. Trial types will be randomized and split so that 80 % of the trials are go trials and 20 % of the trials are no-go trials. The exact timing of stimulus presentation will be determined during pilot testing, but warnings and notifications will be displayed for 500–1000 ms with a jittered inter-stimulus interval of 1000–2000 ms. All stimuli will be normalized in size along the major axis of the image to 750 pixels.

A structural scan will be acquired for each subject for functional localization after which the behavioral task will be completed over two functional runs, each lasting about 8 min. Trials will be randomly assigned to the runs.

Three regions of interest will be examined, namely the visual cortex and ventral visual pathway, medial temporal lobe, and lateral prefrontal cortex. The analysis of each of these regions allows for an examination of generalization at different levels of processing. First, the visual cortex and ventral visual pathway are involved in object perception (see [22] for review). Second, the medial temporal lobe is involved memory specificity, or the detection of differences between similar stimuli [23]. Lastly, the dorso and ventrolateral prefrontal cortices, are involved in response inhibition [24].

3.3 Analysis

The analysis of these different levels of processing allows for the detection of the specific aspect of the stimuli are allowing for greater ability to recognize differences and withhold a response. For instance, if habituation to notifications does not influence generalization to warnings, one would expect to see a greater amplitude of fMRI activation for each warning because it is treated as a new stimulus. If, however, habituation to notifications does generalize to warnings, there is no reason why there would be a greater fMRI activation for warnings over notifications. A warning would be treated as if it was another notification and the habituation would continue. For example, no difference in fMRI activation for the notification and warning trials in the medial temporal lobe might suggest that the notifications and warnings are too similar and the visual design should be changed to make them easier to tell apart. Conversely, if there are differences in the medial temporal lobe, but no differences are observed in the lateral prefrontal cortex, then the problem doesn't lie in the visual differences of the stimulus. Rather the difficulty comes in inhibiting the common response and different designs in the mode of dismissal of the warning or notification (e.g., dragging a bar across the window for a warning rather than clicking a button for a notification) may be better design changes to help prevent this generalization.

4 Anticipated Contributions

We anticipate that our findings of this proposed study will complement and extend previous work that examined habituation to individual warnings. With proposed experimental design, we intend to examine how the effects of habituation to frequent software notifications generalize to novel security warnings. Specifically, our anticipated contributions are:

1. Determine how the effects of habituation to frequent notifications and warnings generalize to novel warnings.
2. Measure the severity of generalization.
3. In a future behavioral study based on these outcomes, develop and test distinctive warning designs that resist generalization.

Acknowledgements This research was funded by a supplement to NSF Grant #CNS-1422831 and a Google Faculty Research Award.

References

1. Bravo-Lillo, C., Cranor, L., Komanduri, S., Schechter, S., Sleeper, M.: Harder to Ignore? Revisiting Pop-up Fatigue and Approaches to Prevent It, USENIX Association, pp. 105–111 (2014)
2. Kalsher, M., Williams, K.: Behavioral compliance: theory, methodology, and result. In: Wogalter, M.S. (ed.) Handbook of Warnings, pp. 313–331. Lawrence Erlbaum Associates, Mahwah, NJ (2006)
3. Rankin, C.H., Abrams, T., Barry, R.J., Bhatnagar, S., Clayton, D.F., Colombo, J., Coppola, G., Geyer, M.A., Glanzman, D.L., Marsland, S., McSweeney, F.K., Wilson, D.A., Wu, C.-F., Thompson, R.F.: Habituation revisited: an updated and revised description of the behavioral characteristics of habituation. Neurobiol. Learn. Mem. **92**(2), 135–138 (2009)
4. Anderson, B.B., Kirwan, C.B., Jenkins, J.L., Eargle, D., Howard, S. Vance, A.: How polymorphic warnings reduce habituation in the brain: insights from an fMRI study. In Proceedings of the 33rd Annual ACM Conference on Human Factors in Computing Systems, ACM, Seoul, Republic of Korea, pp. 2883–2892 (2015)
5. Bravo-Lillo, C., Komanduri, S., Cranor, L.F., Reeder, R.W., Sleeper, M., Downs, J. Schechter, S.: Your attention please: designing security-decision UIs to make genuine risks harder to ignore. In: Proceedings of the Ninth Symposium on Usable Privacy and Security ACM, Newcastle, United Kingdom, pp. 1–12 (2013)
6. Egelman, S., Cranor, L.F. Hong, J.: You've been warned: an empirical study of the effectiveness of web browser phishing warnings. In: Proceedings of the SIGCHI Conference on Human Factors in Computing Systems, ACM, Florence, Italy, pp. 1065–1074 (2008)
7. Thompson, R.F., Spencer, W.A.: Habituation: a model phenomenon for the study of neuronal substrates of behavior. Psychol. Rev. **73**(1), 16–43 (1966)

8. Ramaswami, M.: Network plasticity in adaptive filtering and behavioral habituation. Neuron **82**(6), 1216–1229 (2014)
9. Rankin, C.H.: Introduction to special issue of neurobiology of learning and memory on habituation. Neurobiol. Learn. Mem. **92**(2), 125–126 (2009)
10. Colombo, J., Mitchell, D.W.: Infant visual habituation. Neurobiol. Learn. Mem. **92**(2), 225–234 (2009)
11. Grizzard, M., Tamborini, R., Sherry, J.L., Weber, R., Prabhu, S., Hahn, L., Idzik, P.: The thrill is gone, but you might not know: habituation and generalization of biophysiological and self-reported arousal responses to video games. Commun. Monogr. **82**(1), 64–87 (2015)
12. Simpson, N.: Managing the use of style guides in an organisational setting: practical lessons in ensuring UI consistency. Interact. Comput. **11**(3), 323–351 (1999)
13. Ryan, J.D., Althoff, R.R., Whitlow, S., Cohen, N.J.: Amnesia is a deficit in relational memory. Psychol. Sci. **11**(6), 454–461 (2000)
14. Sunshine, J., Egelman, S., Almuhimedi, H., Atri, N. Cranor, L.F.: Crying Wolf: An Empirical Study of SSL Warning Effectiveness, USENIX Association, pp. 399–416 (2009)
15. Mark, G., Gudith, D., Klocke, U.: The cost of interrupted work: more speed and stress. In: Proceedings of the SIGCHI Conference on Human Factors in Computing Systems, ACM, Florence, Italy, pp. 107–110 (2008)
16. Mark, G., Voida, S., Cardello, A.: A pace not dictated by electrons: an empirical study of work without email. In: Proceedings of the SIGCHI Conference on Human Factors in Computing Systems, ACM, Austin, TX, USA, pp. 555–564 (2012)
17. Amer, T.S., Maris, J.-M.B.: Signal words and signal icons in application control and information technology exception messages—hazard matching and habituation effects. J. Inf. Syst. **21** (2), 1–25 (2007)
18. Anderson, B., Kirwan, B., Eargle, D., Howard, S., Vance, A.: How polymorphic warnings reduce habituation in the brain: insights from an fMRI study. In: Proceedings of the ACM Conference on Human Factors in Computing Systems (CHI), ACM, Seoul, South Korea (2015)
19. Groves, P.M., Thompson, R.F.: Habituation: a dual-process theory. Psychol. Rev. **77**, 419–450 (1970)
20. Riedl, R., Léger, P.-M.: Fundamentals of NeuroIS: information systems and the brain. In: Reuter, M., Montag, C. (eds.) Studies in Neuroscience, Psychology and Behavioral Economics. Springer, Berlin (2016)
21. Nosek, B.A., Banaji, M.R.: The Go/No-Go association task. Soc. Cogn. **19**(6), 625–666 (2001)
22. Grill-Spector, K.: The neural basis of object perception. Curr. Opin. Neurobiol. **13**(2), 159–166 (2003)
23. Kirwan, C.B., Stark, C.E.L.: Overcoming interference: an fMRI investigation of pattern separation in the medial temporal lobe. Learn. Memory. **14**(9), 625–633 (2007)
24. Liddle, P.F., Kiehl, K.A., Smith, A.M.: Event-related fMRI study of response inhibition. Hum. Brain Mapp. **12**(2), 100–109 (2001)

Combating the Influence of the Heuristic Thinking in Online Star Ratings: Preliminary Evidence

Markus Weinmann, Christoph Schneider, and Jan vom Brocke

Abstract Online reviews, often considered more credible and less biased than marketing information, have become an important aspect of making purchase decisions. Yet, online star ratings can be affected by reviewers' heuristic evaluations, potentially leading to suboptimal purchase decisions. For example, star ratings may be biased due to the availability heuristic, i.e., users giving disproportionate weight to one—often negative—attribute. As research has demonstrated that even minor modifications of the presentation of options can have a large influence on people's behavior, we test the effects of prior attribute rating on overall star rating. An experiment ($n = 56$) conducted in the context of restaurant ratings showed that merely asking people first to rate individual attributes can significantly influence overall ratings. These findings can have important implications, as uncovering the effects specific design parameters of review forms have on people's evaluation results will allow for reducing unintended biases in review form design.

Keywords Online judgment • Availability heuristic • Customer ratings

1 Introduction

Electronic word of mouth has become an important aspect of consumer decision making (e.g., [1]), especially in situations where customers cannot experience a service or evaluate a product before making a purchase decision [2]. Often, electronic word of mouth is spread through posts in social media, such as social networks and discussion forums; in addition, online consumers post their evaluations on online retailers' Web sites or dedicated review sites (such as the travel review site tripadvisor.com).

M. Weinmann (✉) • J. vom Brocke
University of Liechtenstein, Vaduz, Liechtenstein
e-mail: markus.weinmann@uni.li; jan.vom.brocke@uni.li

C. Schneider
City University of Hong Kong, Kowloon, Hong Kong
e-mail: christoph.schneider@cityu.edu.hk

F.D. Davis et al. (eds.), *Information Systems and Neuroscience*, Lecture Notes in Information Systems and Organisation 16, DOI 10.1007/978-3-319-41402-7_7

Such product or service reviews are often considered more credible than marketing information [3], but can be subject to various biases (e.g., [4]). Whereas some reviews are missleading due to fraudulent behavior, reviewers' heuristic thinking is also likely to bias reviews. These biases, happening largely subconsciously, can affect a wide range of reviews. In other words, not only the judgments of the product or service but cognitive biases during the writing of the review influence the rating or review content [5].

Literature from psychology and behavioral economics suggests that it may be possible to overcome cognitive biases such as the availability of thoughts [6]. In this paper, we argue that the design of online review forms can help to "debias" users' inputs by reducing potential biases caused by the availability heuristic.

2 Hypothesis

A long line of research has demonstrated that human decision making and judgment is not always rational [7]. As people's cognitive resources are limited, they tend to employ intuitive, heuristic thinking, rather than controlled, analytical reasoning [6, 8]. Studies in the area of cognitive neuroscience have supported this notion, and employed neuroimaging techniques to demonstrate that there are different pathways used in deductive reasoning, with different activations depending on whether people use their heuristic or analytic processing system (e.g., [9]). Likewise, De Neys and Goel [10] used fMRI data to demonstrate differences in activation during decision-making tasks. One such heuristic that can be an important factor influencing people's evaluations of products or services is the availability heuristic, which suggests that the availability of thoughts, or "the ease with which instances [i.e., thoughts of products or services] can be brought to mind" ([6], p. 1127). Relatedly, research has demonstrated that people use "ease of retrieval" as information, retrieve attributes or events more easily from memory, and regard them as more likely or informative [11–13]. Therefore, as people tend to employ heuristic thinking (i.e., heuristic thinking is less effortful than analytical reasoning), overall ratings of products or services are likely to be disproportionally influenced by prevailing positive or negative thoughts. We speculate that form granularity (i.e. presenting several attribute rating dimensions instead of just one overall dimension) can reduce biased judgments caused by the availability heuristic, as asking people to provide detailed attribute ratings is likely to encourage more controlled reasoning and divert attention away from the most easily accessible—often negative—thoughts. As a result, an user interface encouraging attribute ratings (that reminds people about other—more neutral or positive—dimensions) will lead to less negative overall ratings. Thus, we hypothesize that *form granularity will reduce the effect of availability of thoughts*.

3 Methodology

We conducted an online experiment to test the hypothesis. In particular, we asked the participants to read a scenario about a fictitious restaurant visit and rate the restaurant. We randomly assigned the participants to one of two conditions, differing in form granularity (see next paragraph), and analyzed the overall ratings of the restaurant.

3.1 Participants and Experimental Design

We recruited 60 participants from English-speaking countries (USA, Canada, England, Australia, and New Zealand) using Prolific Academic.[1] All participants were paid £1 for a 10-min task (equaling a £6 hourly wage). We eliminated participants who did not read the instructions and excluded anyone who took the experiment on a mobile device from the analysis, resulting in a usable sample size of 56. The experimental design was a single-factor design, with one treatment (prior attribute rating: yes vs. no). We randomly assigned subjects to treatment and baseline groups (treated: 33; baseline: 23). The mean age was 28.84 years, and 45 % were women.

3.2 Materials and Procedure

Scenario Development We developed a fictitious restaurant visit scenario based on the DINESERV instrument, a 29-item scale for measuring consumer perceptions of service quality in restaurants [15], which is based on the SERVQUAL instrument [16]. DINESERV assesses the five dimensions *reliability*, *assurance*, *responsiveness*, *tangibles*, and *empathy*. Using the DINESERV items, we created a story of a fictitious restaurant visit, which we framed as a positive experience.[2] To test the effects of eliciting attribute ratings, we framed the *tangibles* dimension as negative, and all other attributes as positive (this also served as a manipulation check, allowing us to check whether the participants really read the scenario).

Procedure At the beginning of the experiment, participants were told that they would be presented with a scenario of a fictitious restaurant visit that they should rate on a Web site. After reading the scenario, the participants watched a 2-min

[1]Prolific Academic (www.prolific.ac) is an online crowdsourcing market for scientific studies, providing access to a diverse participant pool, and allowing to collect data that have been shown to be as reliable as those collected using other methods [14].

[2]The scenario and measurement items are available upon request.

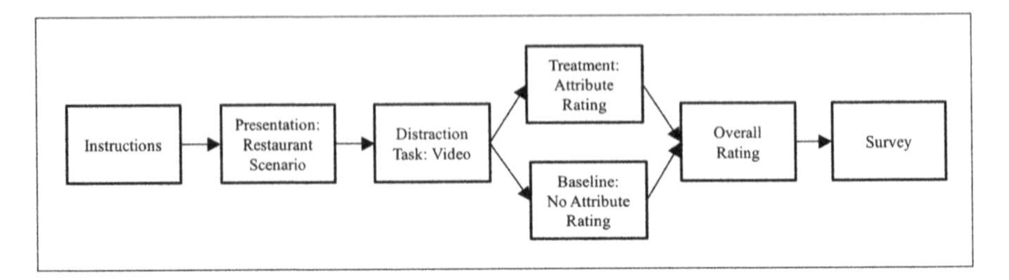

Fig. 1 Procedure of the experiment

animated martial arts film as a distractor task, before rating the restaurant and completing a follow-up survey with demographic questions (see Fig. 1 for the procedure).

Treatment Before asking for an overall rating, subjects in the treatment group were asked to rate the restaurant visit on the attribute scales *service*, *food*, *value*, and *atmosphere* (which are the attributes used on Tripadvisor.com).

3.3 Measures

We asked the participants in the treatment group to provide a detailed rating of the fictitious restaurant visit; in particular, we asked them to rate the dimensions *service*, *food*, *value*, and *atmosphere* on a 9-star scale (participants in the baseline group were not asked to rate these attributes). We then asked all participants to provide an overall star rating of the fictitious restaurant visit on a 9-star scale. As a manipulation check, we asked the participants to rate the service quality of the fictitious restaurant visit using items based on the DINESERV dimensions, as well as the food quality on a 5-point Likert-type scale anchored at very dissatisfied/very satisfied.

4 Results

4.1 Manipulation Check

Using the DINSERV dimensions, we expected *reliability*, *assurance*, *responsiveness*, and *empathy* to be positive as we framed them positive (i.e., high star ratings), but *tangibles* to be negative (low star ratings). Table 1 shows the mean attribute ratings. Whereas all dimensions were rated on average at 4.5 stars (out of 5), *tangibles* was rated on average 2.2, indicating a successful manipulation.

Table 1 Manipulation check whether "tangibles" differ from the other attributes

Attributes	Mean	St. Dev.
Reliability	4.45	0.74
Assurance	4.71	0.49
Responsiveness	4.82	0.43
Tangibles	2.18	0.90
Empathy	4.54	0.57
Food	4.61	0.59

4.2 *Findings*

Mean overall rating in the attribute-rating condition was 6.76 (SD = 0.97) vs. 5.91 (SD = 0.95) in the baseline group, respectively. We used an OLS regression to analyze the data, with our treatment (i.e. *Attribute Rating*) coded as a dummy variable. We specified the following model:

$$Overall\ Rating_i = \alpha_i + \beta_1 \cdot Attribute\ \ Rating_i + \gamma' \cdot Controls_i + \varepsilon_i$$

Table 2 shows the results of the regression analysis. As expected, presenting an attribute rating before asking people to provide an overall rating had a significant positive effect on the overall rating ($\beta = 0.85$; $p < 0.01$; Model 1). The effect was robust when adding control variables ($\beta = 0.86$; $p < 0.01$; Model 2).

5 Discussion and Conclusion

These results show that simple modifications of the design of review forms influence people's evaluation results regarding products and services. More specifically, the results demonstrate that reminding people of evaluation criteria significantly influences overall rating. In other words, asking users of online review sites to rate attributes before providing an overall evaluation of a product or service is likely to make these attributes more salient, and thus influences (or even debiases) the overall rating. These findings support our hypothesis that even minor changes to the design of a rating interface can significantly influence ratings. As such, these findings provide a first step for examining the efficacy of form design in mitigating cognitive biases. Future research may extend these preliminary results by adding further control variables, testing new contexts, and extending the sample size. In addition, future research may use neurophysiological methods, such as capturing and analyzing the participants' gaze fixation, to gain deeper insights and provide a stronger test of the hypothesized relationships [17, 18]. Likewise, research in the context of decision making has shown that the analysis of event-related potentials can be used to study individual differences in susceptibility to different heuristics or biases (e.g., [19–21]); accordingly, EEG may be used to detect the role of biases in the formation of biases and the mechanisms behind the debiasing process. Finally, a

Table 2 Results of the regression analysis. DV: Overall Rating (1–9 stars)

	Model 1: no controls		Model 2: with controls	
Attribute Rating (yes = 1)	0.85***	(0.26)	0.86***	(0.23)
Age			0.05***	(0.01)
Gender			0.43*	(0.23)
Intercept	5.91***	(0.20)	4.24***	(0.46)
Observations	56		56	
R^2	0.16		0.36	
Adj. R^2	0.15		0.32	
F Statistic	10.46*** (df = 1; 54)		8.35*** (df = 3; 52)	

Note: * $p < 0.1$; ** $p < 0.05$; *** $p < 0.01$; standard errors in parentheses

logical extension would be to not only test the effects of attribute rating on star rating but on text rating, e.g., by using text-mining approaches such as latent sentiment analysis (LSA; [22]).

This work has important practical implications: Given the high influence customer reviews have on purchase decisions, the quality of reviews has an important economical and societal impact. We show that threats to review quality do not only stem from deliberate manipulation. Rather, cognitive biases—often happening subconsciously and driven by the way a review form is presented—play an important role. In other words, web designers influence people's online judgments, and, thus, people's choice of products and services. Uncovering the effects specific design parameters of review forms have on people's evaluations may help to reduce unintended biases. For instance, as presented in this study, reminding reviewers of relevant criteria could help to obtain a more balanced evaluation; or, the review form may be designed in a way to intentionally elicit gut feelings by not making use of reminding reviewers on certain evaluation criteria. Research might reveal different types of reviews for which—according to the specific intent of online vendors (i.e., eliciting gut feelings or providing balanced ratings)—different principles of form design may be relevant. Differentiating types of reviews might also help readers of reviews to better interpret them, e.g. is a review a spontaneous snapshots of predominant impressions or rather a balanced and comprehensive evaluation?

Acknowledgements This study has been supported by research grants from the University of Liechtenstein (Project No. wi-2-14) and City University of Hong Kong (Project No. 7004563).

References

1. Mudambi, S.M., Schuff, D.: What makes a helpful online review? A study of customer reviews on Amazon.com. MIS Q. **34**(1), 185–200 (2010)
2. Dellarocas, C.: The digitization of word of mouth: promise and challenges of online feedback mechanisms. Manag. Sci. **49**(10), 1407–1424 (2003)

3. Bickart, B., Schindler, R.M.: Internet forums as influential sources of consumer information. J. Interact. Mark. **15**(3), 31–40 (2001)
4. Muchnik, L., Aral, S., Taylor, S.J.: Social influence bias: a randomized experiment. Science **341**(6146), 647–651 (2013)
5. Schneider, C., Weinmann, M., vom Brocke, J.: Choice architecture: using fixation patterns to analyze the effects of form design on cognitive biases. In: Davis, F.D., Riedl, R., vom Brocke, J., Léger, P.-M., Randolph, A.B. (eds.) Information Systems and Neuroscience. Lecture Notes in Information Systems and Organisation, pp. 91–97. Springer, Switzerland (2015)
6. Tversky, A., Kahneman, D.: Judgment under uncertainty: heuristics and biases. Science **185** (4157), 1124–1131 (1974)
7. Simon, H.A.: A Behavioral model of rational choice. Q. J. Econ. **69**, 99–118 (1955)
8. Goel, V.: Cognitive neuroscience of thinking. In: Berntson, G.G., Cacioppo, J.T. (eds.) Handbook of Neuroscience for the Behavioral Sciences. Wiley, Hoboken, NJ (2009)
9. Goel, V., Dolan, R.J.: Explaining modulation of reasoning by belief. Cognition **87**(1), B11–B22 (2003)
10. De Neys, W., Goel, V.: Heuristics and biases in the brain: dual neural pathways for decision making. In: Vartanian, O., Mandel, D.R. (eds.) Neuroscience of Decision Making, pp. 125–141. Psychology Press, New York, NY (2011)
11. Sherman, S.J., Crawford, M.T., Hamilton, D.L., Garcia-Marques, L.: Social inference and social memory: the interplay between systems. In: Hogg, M.A., Cooper, J. (eds.) The SAGE Handbook of Social Psychology, pp. 45–67. SAGE, London (2007)
12. Schwarz, N., Bless, H., Strack, F., et al.: Ease of retrieval as information: another look at the availability heuristic. J. Pers. Soc. Psychol. **61**(2), 195–202 (1991)
13. Lichtenstein, S., Slovic, P., Fischhoff, B., Layman, M., Combs, B.: Judged frequency of lethal events. J. Exp. Psychol. [Hum. Learn.] **4**(6), 551 (1978)
14. Steelman, Z.R., Hammer, B.I., Limayem, M.: Data collection in the digital age: innovative alternatives to student samples. MIS Q. **38**(2), 355–378 (2014)
15. Stevens, P., Knutson, B., Patton, M.: Dineserv: a tool for measuring service quality in restaurants. Cornell Hotel Restaur. Adm. Q. **36**(2), 5–60 (1995)
16. Parasuraman, A., Zeithaml, V.A., Berry, L.L.: SERVQUAL: a multiple-item scale for measuring consumer perceptions of service quality. J. Retail. **64**(1), 12–40 (1988)
17. Dimoka, A., Banker, R.D., Benbasat, I., et al.: On the use of neurophysiological tools in is research: developing a research agenda for NeuroIS. MIS Q. **36**(3), 679–702 (2012)
18. Riedl, R., Banker, R., Benbasat, I., et al.: On the foundations of NeuroIS: Reflections on the Gmunden Retreat 2009. Commun. Assoc. Inf. Syst. **27**(1), Art 15 (2010)
19. De Neys, W., Novitskiy, N., Ramautar, J., Wagemans, J.: What makes a good reasoner?: Brain potentials and heuristic bias susceptibility. In: Proceedings of the 32nd Annual Conference of the Cognitive Science Society (2010)
20. Ma, Q., Li, D., Shen, Q., Qiu, W.: Anchors as semantic primes in value construction: an EEG study of the anchoring effect. PloS One **10**(10), e0139954 (2015)
21. Schwikert, S.R., Curran, T.: Familiarity and recollection in heuristic decision making. J. Exp. Psychol. Gen. **143**(6), 2341 (2014)
22. Deerwester, S.C., Dumais, S.T., Landauer, T.K., Furnas, G.W., Harshman, R.A.: Indexing by latent semantic analysis. J. Am. Soc. Inf. Sci. **41**(6), 391–407 (1990)

Mobile BCI Technology: NeuroIS Goes Out of the Lab, into the Field

S.C. Wriessnegger, G. Krumpl, A. Pinegger, and G.R. Mueller-Putz

Abstract In the past years many NeuroIS studies have been published using different neuroimaging tools like electroencephalography (EEG) or functional magnetic resonance imaging (fMRI). In general most of the EEG studies have been performed in the lab, where participants are mounted with EEG sensors sitting in front of the computer and following the presented instructions. There are several mobile EEG systems on the market which could be used to investigate brain activity of human behaviour in the field, like during sports or social activities. In this paper we will present a novel system for EEG-based NeuroIS studies out of the lab, named mobile NeuroIS. The system consists of a wireless EEG system and a smartphone, serving as a monitor. Beside the system architecture we will present first evaluation data of three participants using it as mobile Brain-Computer Interface (BCI) application.

Keywords Mobile EEG • P300 speller • Brain-computer interfaces

1 Introduction

Small and wireless mobile EEG systems have substantial practical advantages as they allow for brain activity recordings in natural environments [1–3]. This technology facilitates among others the transfer of brain-computer interface (BCI) applications from the laboratory to natural daily life environments, one of the key challenges in current BCI research [4].

A recent definition describes a BCI as follows [5, 6] "A BCI is a system that measures central nervous system (CNS) activity and converts it into artificial output that replaces, restores, enhances, supplements, or improves natural CNS output and thereby changes the ongoing interactions between the CNS and its external or internal environment." Recently a Special Issue called "The plurality of Human Brain-Computer Interfacing was published, delivering the state of the art in BCI research from applications [7] to signal processing [8]."

S.C. Wriessnegger (✉) • G. Krumpl • A. Pinegger • G.R. Mueller-Putz
Institute of Neural Engineering, Graz University of Technology, Graz, Austria
e-mail: s.wriessnegger@tugraz.at

F.D. Davis et al. (eds.), *Information Systems and Neuroscience*, Lecture Notes in Information Systems and Organisation 16, DOI 10.1007/978-3-319-41402-7_8

In the past years a shift of interest in BCI technology comes from the unique advantage of having access to the user's ongoing brain activity [9] which enables applications spanning a variety of domains such as brain-activity based gaming [10] or neuro-economics [11]. For example, Misawa and colleagues created a trial BCI system to assist with purchase decision-making. Furthermore [12] already outlined two possible long-term goals of BCI research in the business domain: (1) automatization of process steps in administrative work flows (e.g., enterprise systems recognize a user's thoughts and start with information processing without using an input device), and (2) increase of a system's usability (e.g., automatic redesign of an interface on the basis of a user's mental state).

Here we will present a mobile NeuroIS system, consisting of a mobile EEG system and a standard mobile phone and its evaluation with data of a classical P300-based BCI task. For example De Vos and colleagues [4] compared their mobile EEG [2] to a well-established, wired laboratory EEG amplifier. They also used a classical visual ERP speller paradigm [13], as it is among the most frequently used BCI paradigms. Compared to our study, where the participants performed the experimental task on the smartphone, they were sitting in front of a computer in the lab running the experiment. Especially EEG, with a high temporal resolution, became a tool of growing interest for investigations in IS research [14]. In the past years a lot of IS constructs have been investigated by EEG but mostly restricted to laboratory situations. Therefore our proposed system can also be used in future NeuroIS research in natural environments.

2 System Architecture

The mobile P300 speller was implemented in an android-based biosignal acquisition application called mobile BCI. This application allows transformation of EEG data through a given network. It connects via Bluetooth 2.0 the biosignal amplifier. For this purpose the TOBI Signalserver and TiA interface [15] was migrated in a mobile application and a user interface for controlling was implemented. In the following part the method to develop a P300 speller for android based devices will be explained in detail.

2.1 Implementation General

The implementation was realized in JAVA by using the Android Software Development Kit (SDK). This toolkit delivers only the basic functions. Additionally tools like the Android platform tools and the Android library package are necessary to implement a user-friendly design. As integrated development environment (IDE), Eclipse with the Android Development Tools (ADT) Plugin was used. The ADT Plugin allows successful development of an Android Applications.

2.2 Client

The client receives and saves data of the Signalserver for further processing. The Signalserver used the TOBI Interface A (TiA) library which is implemented in C++ to support a standardized data transfer. Using the TiA for the current application the Android native Development Kit (NDK) was necessary. This NDK is able to compile the source code (C, C++, etc.) in native code libraries or executables. For this purpose a Client.java class was implemented which implements the compiled shared library with the loadLibrary() method. With the IP address and the port number it allows setting up or quit a connection to the Signalserver by TCP or UDP.

2.3 User Interface

This section describes the realisation of the P300 speller, the online feedback and related options. The 6×6 speller Matrix was improved by using a SurfaceView. Especially when fast updates are necessary the SurfaceView is essential. To build a SurfaceView a new class (SpellerSurfaceView.java) was established and the SurfaceHolder.Callback implemented. The SurfaceHolder.Callback Interface informs the SurfaceView if the surface has been build, changed or destroyed. The original 6×6 matrix developed by [16] was enhanced with 36 alphanumeric signs. Furthermore a Textbox and a Textview were developed to give the user feedback and display the current status respectively. Additionally the options menu of the speller was improved to control several settings for different operation modes.

3 Pilot Study

Our system constitutes a fully portable EEG based real-time system including stimulus delivery, data acquisition, brain state decoding and neurofeedback. The raw EEG data is acquired with a mobile wireless biosignal acquisition system (g.MOBIlab, g.tec, Austria) to record the EEG and a common smartphone (Fig. 1). The system works with each commercially available smartphone after successful installation of the mobile BCI App.

The EEG data were transmitted to a receiver module connected to the smartphone. Custom-made software (TOBI signalserver and TiA interface [15]) for the phone transmits the EEG data to a server or processes it locally, enabling real-time brain state decoding. In the following section we describe the evaluation of this system with three volunteers performing a traditional P300 based BCI task.

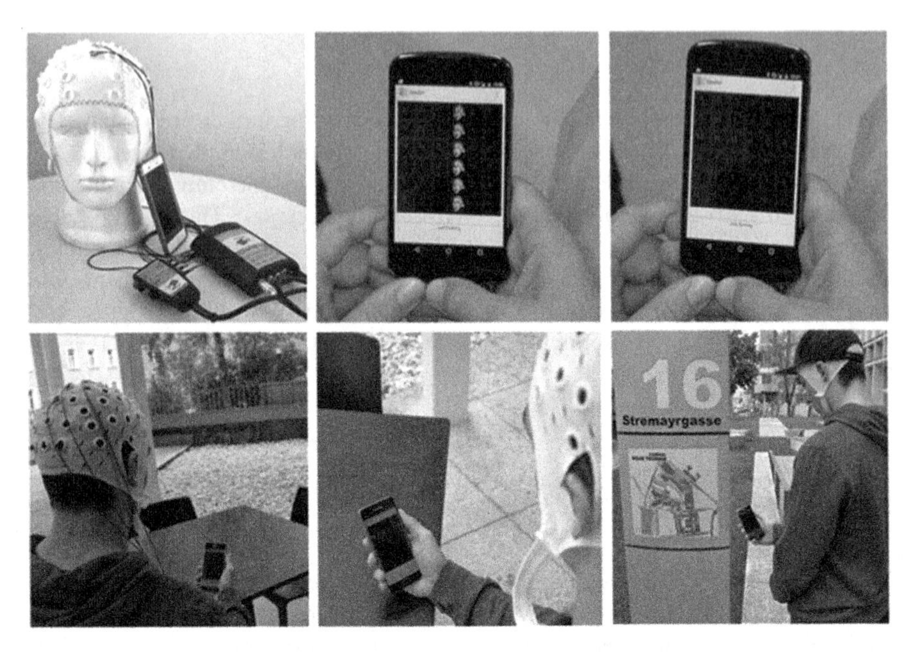

Fig. 1 *Top left*: System setup with commercial smartphone. *Right*: Smartphone display with flashing and not flashing speller matrix; *Bottom*: Participant using P300 Speller out of the lab

4 Methods

4.1 Participants

Three participants, one female, two male, aged between 19 and 30 took part in the pilot study. All participants received a detailed explanation of the experimental procedure and written informed consent was obtained from all. The study was carried out in accordance with The Code of Ethics of the World Medical Association (Declaration of Helsinki) for experiments involving humans.

4.2 EEG Recording

In the present study we used a mobile wireless biosignal acquisition System (g.MOBllab, g.tec, Austria) to record the EEG. We mounted four passive Ag/AgCl (Fz, Cz, Pz, Oz) electrodes according to the 10/20 international system. A sampling rate of 256 Hz and a bandpass filter between 0.5 and 100 Hz was used. It is also possible to use other EEG sensors [17].

4.3 Paradigm

The P300 speller paradigm was used, where the commands (letters or symbols) are displayed in a matrix on a screen. When the user focuses on an element of the matrix a P300 and other event-related potentials (ERPs) are elicited whenever this element is highlighted. The rows and columns are then highlighted by chance and the one containing the target item will elicit the P300 response. Using this procedure, a visually evoked P300 is used to drive a communication system that is usually referred to as "P300 speller" [18, 19]. This "P300 speller" typically consists of a keyboard-like matrix, containing alphanumeric characters which are usually presented on a personal computer screen to the user [20–22]. Here we use the smartphone display as the screen enabling the participant to use the P300 speller everywhere out of the lab (Fig. 1, bottom).

4.4 Training

In a first session the calibration of the system was realised by copy spelling of the defined letters/numbers "H4FUJ". After this calibration of the classifier a rest period of 10 min followed. The EEG data of participant 1 for the calibration session are illustrated in Fig. 2 exemplarily.

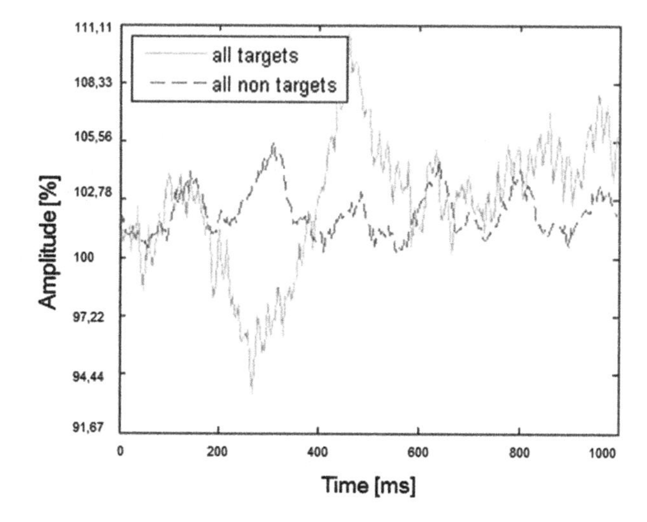

Fig. 2 Grand average (all letters) of participant 1

Table 1 Accuracy [%] and bit-rate [bits/min] of each online spelling session

	Participant 1		Participant 2		Participant 3	
Trial	Accuracy	Bit-rate	Accuracy	Bit-rate	Accuracy	Bit-rate
1	93.33	6.40	100.00	6.40	94.44	7.04
2	86.66	5.49	93.33	6.40	88.88	6.23
3	100.00	7.57	93.33	7.57	94.44	7.04
Mean/SD	**93.33/6.67**	**6.49/1.04**	**95.55/3.85**	**6.79/0.86**	**92.58/3.21**	**6.77/0.49**

4.5 Online Spelling

After the training session and system calibration participants had to spell "MOBILE BCI TEST" three times. During this spelling session participants received feedback immediately when the desired letter appeared. In case of wrong selections they were instructed to keep on spelling. The results of the online spelling task are shown in Table 1. For each trial of the online spelling task and participant, the accuracy and the bitrate [bits/min] are shown in Table 1. All three participants achieved accuracies above 85 % in all trials reflecting a very good performance.

Concluding, the results of our pilot study showed that the system works properly and could be used on a commercial smartphone. A future study is planned with more participants and a modified P300 spelling setup and application.

5 Discussion

In the present paper we introduced a mobile EEG system with a smartphone serving as monitor and recording device. The simple system architecture allows applications in natural environments, out of the lab. The system could be easily adapted to other experimental designs and also other sensor types, e.g. active or dry electrodes, could be used. We evaluated the system with the P300 speller and the results are very promising. The mean accuracy was in all participants higher than 90 % with a mean bit-rate of about 6 s. This first pilot study shows the suitability of the proposed EEG system for real-time neuroscience applications, based on standard mobile phone technology. Furthermore it can easily be applied to different experimental scenarios investigating constructs like technostress or decision making in the field of NeuroIS.

References

1. Kranczioch, C., Zich, C., Schierholz, I., Sterr, A.: Mobile EEG and its potential to promote the theory and application of imagery-based motor rehabilitation. Int. J. Psychophysiol. **91**(1), 10–15 (2014). ISSN 0167-8760

2. Debener, S., Minow, F., Emkes, R., Gandras, K., De Vos, M.: How about taking a low-cost, small, and wireless EEG for a walk? Psychophysiology **49**, 1449–1453 (2012)
3. Park, J.L., Fairweather, M.M., Donaldson, D.I.: Making the case for mobile cognition: EEG and sports performance. Neurosci. Biobehav. Rev. **52**, 117–130 (2015)
4. De Vos, M., Kroesen, M., Emkes, R., Debener, S.: P300 speller BCI with a mobile EEG system: comparison to a traditional amplifier. J. Neural Eng. **11**, 3 (2014)
5. Wolpaw, J., Winter Wolpaw, E.: Brain-computer interfaces: something new under the sun. In: Wolpaw, J., Winter Wolpaw, E. (eds.) Brain-Computer Interfaces: Principles and Practice, S. 3–12. Oxford University Press, New York (2012)
6. Brunner, C., Birbaumer, N., Blankertz, B., Guger, C., Kübler, A., Mattia, D., Millán, J.D.R., Miralles, F., Nijholt, A., Opisso, E., Müller-Putz, G.R.: BNCI Horizon 2020: towards a roadmap for the BCI community. Brain-Comput. Interfaces **2**, 1–10 (2015)
7. Müller-Putz, G., et al.: Towards noninvasive hybrid brain-computer interfaces: framework, practice, clinical application, and beyond. Proc. IEEE **103**(6), 926–943 (2015)
8. Lotte, F.: Signal processing approaches to minimize or suppress calibration time in oscillatory activity-based brain-computer interfaces. Proc. IEEE **103**(6), 871–890 (2015)
9. Zander, T., Kothe, C.: Towards passive brain-computer interfaces: applying brain-computer interface technology to human-machine systems in general. J. Neural Eng. **8**, 025005, 1–5 (2011)
10. Nijholt, A.: BCI for games: a 'state of the art' survey. In: ICEC'08: Proceedings of the 7th International Conference on Entertainment Computing, pp. 225–228. Springer, Berlin (2009)
11. Misawa, T., et al.: A brain-computer interface for purchase decision-making. Int. J. Comput. Sci. **4**(2), 173–185 (2010)
12. Riedl, R., Banker, R.D., Benbasat, I., Davis, F.D., Dennis, A.R., Dimoka, A., Gefen, D., Gupta, A., Ischebeck, A., Kenning, P., Müller-Putz, G., Pavlou, P.A., Straub, D.W., vom Brocke, J., Weber, B.: On the foundations of NeuroIS: reflections on the Gmunden Retreat 2009. Commun. AIS **27**, 243–264 (2010)
13. Kübler, A., Kotchoubey, B., Kaiser, J., Wolpaw, J.R., Birbaumer, N.: Brain–computer communication: unlocking the locked in. Psychol. Bull. **127**, 358–375 (2001)
14. Müller-Putz, G., Riedl, R., Wriessnegger, S.: Electroencephalography (EEG) as a research tool in the information systems discipline: foundations, measurement, and applications. Commun. Assoc. Inf. Syst. **37**, 911–948 (2015)
15. Breitweiser, C., Neuper, C., Müller-Putz, G.R.: 2011. TOBI interface A (TiA)—a standardized interface to transmit raw biosignals. Int. J. Bioelectromagn. **13**, 64–65 (2011)
16. Farwell, L., Donchin, E.: Talking the top of your head: toward a mental prosthesis utilizing event-related brain potentials. Electroencephalogr. Clin. Neurophysiol. **70**(6), 510–523 (1988)
17. Stopczynski, A., Larsen, J.E., Stahlhut, C., Petersen, M.K., Hansen, L.K.: A Smartphone interface for a wireless EEG headset with real-time 3D reconstruction. In: D'Mello, S., Graesser, A., Schuller, B., Martin, J.-C. (eds.) Affective Computing and Intelligent Interaction. Lecture Notes in Computer Science, vol. 6975, pp. 317–318. Springer, Berlin (2011)
18. Donchin, E., Spencer, K.M., Wijesinghe, R.: The mental prosthesis: assessing the speed of a P300-based brain-computer interface. IEEE Trans. Neural Syst. Rehabil. Eng. **8**, 174–179 (2000)
19. Aloise, F., et al.: P300-based brain-computer interface for environmental control: an asynchronous approach. J. Neural Eng. **8**(2) (2011)
20. Halder, S., Pinegger, A., Käthner, I., Wriessnegger, S.C., Faller, J., Antunes, J.B.P., Müller-Putz, G.R., Kübler, A.: Brain-controlled applications using dynamic P300 speller matrices. Artif. Intell. Med. **63**(1), 7–17 (2015)
21. Pinegger, A., Faller, J., Halder, S., Wriessnegger, S.C., Müller-Putz, G.R.: Control or non-control state: that is the question! An asynchronous visual P300-based BCI approach. J. Neural Eng. **12**(1), 1–8 (2015)
22. Pinegger, A., Wriessnegger, S.C., Müller-Putz, G.R.: Introduction of a universal P300 brain-computer interface communication system. Biomed. Eng. **58**, 1–2 (2013)

Exploring the Mental Load Associated with Switching Smartphone Operating Systems

Tyron Booyzen, Aaron Marsh, and Adriane B. Randolph

Abstract The increasing importance of the smartphone industry has led many to debate which leading mobile platform, iOS or Android, is best. One of the central arguments is that perceived ease of use of the device should be the principle consideration. To measure this, one can reference Cognitive Load Theory and employ EEGs to register the mental load associated with completing common smartphone tasks such as making a call or navigating to a web page. The proposed study aims to evaluate the mental load associated with a set of tasks performed on a smartphone by a user who is unfamiliar with the operating system. It is our belief that this initial measure of mental load will act as a surrogate for which operating system embodies the idea of universal usability.

Keywords Cognitive load • Mental load • Perceived ease of use • iOS • Android • Implicit measure • Universal usability • Smartphone • NeuroIS

1 Introduction

The continued growth of the smartphone market globally coupled with the fact that almost three quarters of smartphone buyers reportedly care about the operating system (OS) that runs on their device [1] has led to the research question of *which mobile platform, Apple's iOS or Google's Android, is better*? While many reviewers on technology review websites have fought over this question by delineating key differences between the two such as user experience, customizability, reliability, battery life management, and app ecosystems [2], no one has offered conclusive scientific evidence of which one is better. Furthermore, the argument is marred by the intense loyalty that can be observed of both camps [3].

One of the best arguments in this debate is that a smartphone OS should be evaluated by its ease of use to first-time users. While this debate is impelled by technology bloggers the world over, it is founded on academic thought. Fred Davis

T. Booyzen • A. Marsh • A.B. Randolph (✉)
BrainLab, Kennesaw State University, Kennesaw, GA, USA
e-mail: tbooyzen@students.kennesaw.edu; amarsh7@students.kennesaw.edu; arandolph@kennesaw.edu

F.D. Davis et al. (eds.), *Information Systems and Neuroscience*, Lecture Notes in Information Systems and Organisation 16, DOI 10.1007/978-3-319-41402-7_9

[4] began outlining this theory when he suggested that when predicting usability, one should consider its perceived ease of use which he defined as "the degree to which a person believes that using a particular system would be free of effort" (p. 320). He postulated that this was the best measure of how much a system would be used and accepted with the Technology Acceptance Model (TAM). To validate his claims he employed theories of motivation including the theory of self-efficacy [5] which is defined as the "judgments of how well one can execute courses of action required to deal with prospective situations" (p. 122).

Ultimately, an OS and a smartphone are co-designed to help the user accomplish certain tasks: call a contact, add an appointment to their calendar, navigate to a specific web page, check the weather, get directions, etc. These tasks should be completed with ease and without much thought by the user. The smartphone should represent a tool that acts as an extension of the user and should never present a barrier to completing a given task, especially when the user has limited time. This is a concept termed *universal usability* as described by Ben Shneiderman [6].

Universal usability is something that every OS vendor, Apple and Google included, should work towards achieving. However, it is not clear how one would identify universal usability since there is no definitive way to measure it. This is rooted in the fact that perception is unique to every individual and entirely intrinsic; it is difficult to compare between individuals through articulation. Thus, the use of neurophysiological tools to measure universal usability as an outlay of perceived ease of use may hold the answer. Other NeuroIS researchers would seem to agree as elements of TAM have been examined using electroencephalography (EEG) [7] and functional magnetic resonance imaging (fMRI) [8].

In neuroscience and psychology the Cognitive Load Theory (CLT) is based on an understanding that the brain consists of "limited working memory with partly independent processing units for visual/spatial and auditory/verbal information, which interacts with a comparatively unlimited long-term memory" (p. 63) [9]. While this framework has concentrated on the development of instructional methods, it can be applied to the evaluation of difficulty which can be linked to perceived ease of use. Therefore, through the measurement of cognitive load—or more specifically mental load, which is that aspect of cognitive load that is derived from the demands of completing a task—one may effectively measure the perceived ease of use in a particular instance. In this case, the perceived ease of use of an OS.

It is important to note, however, that mental load experienced when completing a task is dramatically reduced through repetition which leads to the bypassing of working memory. As outlined by Charles Duhigg [10], this is the moment that the Basal Ganglia takes over from the neocortex. This can be further explained by the fact that the brain is a very complex organ that acts selfishly seeking to reduce mental load as much as possible [11]. Therefore, users of a particular OS will experience their cognitive load diminishing when completing common tasks as they become accustomed to the OS and its user interface. To truly study the perceived ease of use of a platform, one must eliminate the possibility of habitual operation and focus only on users who are new to an OS.

In measuring the mental load of smartphone users as they switch from a familiar OS to an unfamiliar OS, it is hypothesized that one could effectively evaluate the perceived ease of use of the unfamiliar OS. Furthermore, the mental load experienced when using a familiar OS can be used as a baseline for when the same OS is unfamiliar to a user.

Thus, the purpose of this study is to evaluate Apple's iOS and Google's Android platforms in terms of perceived ease of use, through the measurement of mental load. In doing so, it is believed that a more objective answer can be established to the question of which OS is better. More importantly, however, this will provide an opportunity to both vendors and other user interface and OS designers to evaluate their own performance. The importance of this is greatly emphasized when one considers the link between perceived ease of use and user satisfaction and then user satisfaction and profitability.

2 Background

Ever since the Apple iPhone's original launch in 2007 and Google Android's first appearance in September of 2008, there has existed a feud between the two camps. Similar to the feud between Chevy and Ford trucks or Playstation and Xbox gaming consoles, it appears that a defined winner may never arise. This is due to the fact that both sides offer something that the other does not. This can make for a frustrating, never-ending argument between both software development giants. There are, however, benefits to conflicts such as these. With both sides wanting to win this "war" and gain a larger share of the market, technology within the OS is constantly being refined and enhanced to try and out-do the competitors, which benefits consumers greatly.

In this study, we will utilize a flagship smartphone on the iPhone and the Android OS. It is crucial that the smartphones be of "flagship" status since we want both OSs to run at their optimum level. For iOS, we will use an iPhone 6 s and for Android we are targeting use of a Nexus 6P (manufacturer willing). We chose to use an iPhone 6 s because it is the only smartphone that runs iOS and it is of the current generation of iPhone. Out of all of the available Android devices on the market (e.g., Samsung, LG, HTC, Motorola, etc.), we chose Nexus because it runs the purest form of Android currently on a smartphone platform, according to the Nexus website (https://www.google.com/nexus/). It is important to note that we are not testing hardware strength, but the ease of use of the OS.

Through this study, we will be measuring mental load to determine which of the two leading mobile OSs, iOS or Android, ranks top in perceived ease of use. It is our belief that iOS will come out on top and have lower levels of mental load based on the high levels of satisfaction that users of the platform report [12] and due to its widespread use by consumers. Therefore, the contrarian belief is that Android will cause higher levels of mental load, and subsequently require more time to complete specific tasks than iOS. The purpose of the study, however, is not necessarily to acknowledge a definite "winner," but to better understand how user interfaces help to reduce mental load through the aesthetics and functionality of the OS.

3 Proposed Methodology

A sample of 32 college students aged 20–25 will be drawn randomly from a university located in the southeast of the United States. Sixteen (16) participants will be incumbent Android users and 16 participants will be incumbent iOS users. An incumbent user, for the purposes of this study, is any user that has used a particular operating system for 2 years or more. Participants will be screened to ensure that they do not interact with the 'unfamiliar OS' through another medium such as a tablet.

The participants will then be recorded using EEG to measure their mental load as they complete a set of five simple smartphone tasks: (1) Make a call, (2) Send a text message with a photo, (3) Navigate to a webpage, (4) Set a time-based reminder, and (5) Get directions. These tasks will be performed and the resulting mental load will be recorded for each task in randomized order on a familiar device and on an unfamiliar device. This methodology will follow similarly to that used by de Guinea et al. [7] in measuring mental load as an implicit antecedent of perceived ease of use, including the use of survey questions coupled with qualitative questions to help triangulate perceptions.

The results will be analyzed and each task completed by each participant will be ranked in terms of mental load required. These will then be tallied at the end of the study to determine which OS produces the lower amount of mental load for new users and compared with the qualitative statements made.

4 Conclusion

This study will be testing to see which mobile OS, iOS or Android, causes the least amount of mental load on new users by measuring and recording more objective data in the form of EEG readings. This is not only to demonstrate which mobile OS is universally more compatible with end users, but to observe if the particular design and functionality of a user interface really does play a large part in perceived ease of use. College students will be recruited and have their EEGs recorded to measure their mental load as they complete defined tasks. After data has been recorded and analyzed, we believe that iOS will come out on top due to its high customer satisfaction ratings over its lifetime, and that Android will cause greater mental load for new users, demonstrable via EEG activations.

References

1. Cromar, S.: Smartphones in the US: market analysis. https://www.ideals.illinois.edu/bitstream/handle/2142/18484/Cromar,%20Scott%20-%20U.S.%20Smartphone%20Market%20Report.pdf (2010)
2. Hill, S.: Which smartphone OS wins 2015? Android Marshmallow vs. iOS 9 vs. Windows 10 Mobile. http://www.digitaltrends.com/mobile/best-smartphone-os/2/ (2015)
3. Pandey, M., Nakra, N.: Consumer preference towards smartphone brands, with special reference to android operating system. IUP J. Mark. Manag. **13**(4), 7–22 (2014)
4. Davis, F.D.: Perceived usefulness, perceived ease of use, and user acceptance of information technology. MIS Q. **13**(3), 319–340 (1989)
5. Bandura, A.: Self-efficacy mechanism in human agency. Am. Psychol. **37**(2), 122–147 (1982)
6. Shneiderman, B.: Leonardo's Laptop: Human Needs and the New Computing Technologies. The MIT Press, Cambridge, MA (2002)
7. de Guinea, A.O., Titah, R., Léger, P.-M.: Explicit and implicit antecedents of users' behavioral beliefs in information systems: a neuropsychological investigation. J. Manag. Inf. Syst. **30**(4), 179–210 (2014)
8. Dimoka, A., Davis, F.: Where does TAM reside in the brain? The neural mechanisms underlying technology adoption. In: Proceedings of the 29th International Conference on Information Systems, Paris, pp. 1–18 (2008)
9. Paas, F., Tuovinen, J.E., Tabbers, H., Van Gerven, P.W.: Cognitive load measurement as a means to advance cognitive load theory. Educ. Psychol. **38**(1), 63–71 (2003)
10. Duhigg, C.: The power of habit, why we do what we do in life and business. https://itunes.apple.com/us/book/the-power-of-habit/id446670958?mt=11 (2012)
11. Genco, S.J., Pohlmann, A.P., Steidl, P.: Neuromarketing for dummies. https://itunes.apple.com/us/book/neuromarketing-for-dummies/id718485749?mt=11 (2013)
12. Kahn, J.: Apple tops J.D. Power's 2014 survey in smartphone satisfaction across all U.S. carriers. http://9to5mac.com/2014/04/24/apple-tops-j-d-powers-2014-survey-in-smartphone-satisfaction-across-all-carriers/ (2014)

Samsung Versus Apple: Smartphones and Their Conscious and Non-conscious Affective Impact

Peter Walla and Markus Schweiger

Abstract Traditional market research used to focus on survey-based investigations. The first generation of Neuromarketing focused on Brain Imaging via functional Magnet Resonance Imaging (fMRI), which allowed for a true insight into brain activities with accurate localisation results. Finally, the new generation of Neuromarketing provides valuable information about implicit affective responses, which in combination with explicit responses are most useful for product evaluation and many further economic decisions. This study investigated various aspects of Samsung and Apple smartphones with respect to their conscious and unconscious affective impact in response to visual presentations.

Besides various interesting discrepancies between conscious and unconscious affective responses, which demonstrate that mainly conscious cognitive differences exist between Apple and Samsung, an overall outcome is that male Samsung owners demonstrated most positive nonconscious affective processing levels regardless of which brand and version being exposed to.

Keywords Smartphones • Conscious and unconscious affective responses • Neuromarketing • Consumer neuroscience

P. Walla (✉)
CanBeLab, Department of Psychology, Webster Vienna Private University, Praterstrasse 23, 1020 Vienna, Austria

School of Psychology, Newcastle University, Callaghan, NSW, Australia

Faculty of Psychology, University of Vienna, Vienna, Austria
e-mail: peter.walla@webster.ac.at

M. Schweiger
Faculty of Psychology, Sigmund Freud Private University Vienna, Campus Prater, Freudplatz 1, 1020 Vienna, Austria
e-mail: schweiger.markus@gmail.com

F.D. Davis et al. (eds.), *Information Systems and Neuroscience*, Lecture Notes in Information Systems and Organisation 16, DOI 10.1007/978-3-319-41402-7_10

1 Introduction

Most companies make their profits by selling products that can vary from actual goods to services provided to customers. In any case, products are designed and shaped and once they are readily created they are evaluated for the purpose of testing their usability and aesthetics both of which are meant to be indicators for consumers purchasing behavior. The traditional approach to this is to utilize surveys and/or questionnaires in order to collect explicit responses that are then used to adjust goods or services before they finally hit the market. However, a number of rather recent investigations has highlighted that explicit responses can lead to opposite results compared to implicit responses [1–5], which are often understood as being more involved in the guidance of human behavior than any conscious thinking processes [6].

The current study followed the trend to simultaneously take explicit and implicit measures while participants were exposed to visual presentations of two versions of iPhones (Apple) and two versions of Galaxy phones (Samsung). In addition, the Apple logo was compared to the Samsung logo as well as the iPhone package design to the Galaxy package design with respect to potential discrepancies between implicit and explicit responses with a focus on affective processing.

We used startle reflex modulation (SRM) as an index of implicit (non-conscious) affective responses and subjective rating performance as an index of explicit (conscious) emotion-related processes.

2 Methods

2.1 Participants

Participants were 31 volunteers aged from 19 to 25 (14 men). All were right-handed, with normal or corrected to normal vision and without neuropathological history. They were tertiary education students familiar with both smartphone brands, Apple and Samsung. They all signed their informed consent and the study was approved by a local ethics committee.

2.2 Procedure

Participants were invited to come to the CanBeLab (Cognitive and Affective Neuroscience & Behavior Lab) at the Webster Vienna Private University in order to volunteer in this study. They were given an information statement and a consent form to sign as well as a short demographic questionnaire to fill in. They were then asked to sit on a comfortable chair while sensors for startle eye blink responses, heart rate and skin conductance were applied. All respective data were recorded with a Nexus 10 mobile and wireless physiological recording device (Mind Media Netherlands) by using the associated Biotrace + software package. Biotrace + was also used to control visual stimuli and startle probe presentations (50 ms bursts of acoustic white noise delivered via headphones at a sound pressure level of about 110 dB).

Two presentation blocks were created, one of which included photos of actual iPhones and Galaxy phones, while the other block contained photos of the Apple and Samsung logo plus iPhone and Galaxy package boxes, all presented in random order. For each photo category, 30 different versions were presented and every 5th or 6th presentation was associated with a startle probe resulting in five startle responses per condition and per subject.

2.3 Startle Reflex Modulation and Explicit Ratings

A 10 channel mobile physiological recording device (NeXus-10 by Mind Media BV) in combination with the software package Bio-trace + was used to measure eye blink responses to controlled startle probe presentation. Through bipolar electromyography left eye blink responses that are contractions of the *musculus orbicularis oculi* were measured of every study participant. For a more detailed description of the respective EMG recording technique, please refer to Mavratzakis et al. (2013) [7].

For this study, both heart rate as well as skin conductance data were not further analysed. Thus, only startle reflex modulation is explained here. After each presentation participants had to rate their like and dislike respectively on a Likert scale from 1 to 9 button presses.

2.4 Data Processing and Statistical Analysis

Explicit ratings of all target presentations that had startle probes associated with were averaged across all five separate versions for each condition. In order to analyse startle responses the root mean square (RMS) method was used to transform all raw high frequency EMG signals into amplitudes, which were then taken for descriptive as well as analytic statistics. Averages were created across all five startle responses for each condition.

In the following, Repeated Measures ANOVA and t-tests were calculated to compare means of all measures and all conditions for both explicit rating and startle response data.

3 Results

3.1 Smartphone Package Design

Explicit rating data show that owners of iPhones like Apple packages more than Samsung packages and vice versa that Galaxy phone owners like Samsung packages better than Apple packages (see Table 1: ANOVA interaction p-value = 0.005, f = 6.551) (see Fig. 1 for bar diagrams). Gender did not significantly interact.

The ANOVA analysis of startle responses did result in only a trend towards significance with respect to a three way interaction between *brand*, *owned smartphone* and *sex*. This trend shows that male iPhone owners respond more

Table 1 ANOVA results related to Package design

PACKAGE-valence rating	*F*	*p-value*
Brand	0.010	0.921
Brand*Sex	0.590	0.450
Brand*Owned	6.551	**0.005**
Brand*Sex*Owned	1.268	0.299
PACKAGE-startle reflex modulation	*F*	*p-value*
Brand	0.883	0.356
Brand*Sex	0.723	0.403
Brand*Owned	0.835	0.446
Brand*Sex*Owned	2.471	0.105

Significant value is indicated in Bold

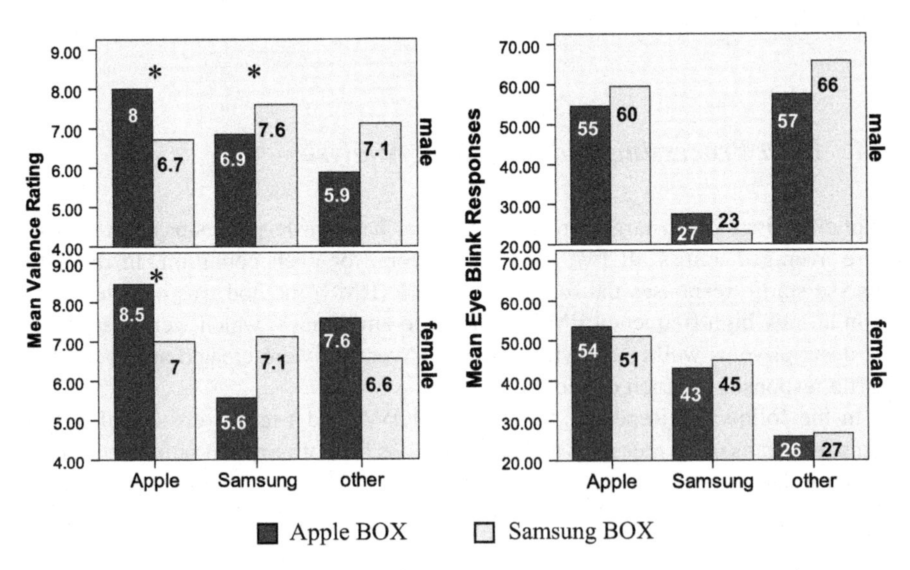

Fig. 1 While t-tests revealed partially significant valence rating differences (*) no startle response differences were found to be significant, although strong trends occurred

positive to iPhone packages and more negative to Galaxy packages, while male Galaxy phone owners respond more positive to Galaxy packages and more negative to iPhone packages (matching result). In females, the effects were reversed, however, this results did not reach significance (see Table 1 and Fig. 1).

3.2 Smartphone Logo

In Apple owners, both sexes rated the Apple logo more positive than the Samsung logo, and even Galaxy owners found the Apple logo more positive than the Samsung logo (less pronounced though). Interestingly, female other smartphone owners rated the Apple logo more positive than the Samsung logo, but in male other smartphone owners this was reversed.

LOGO-valence rating	*F*	*p-value*
Brand	10.564	**0.003**
Brand*Sex	0.456	0.506
Brand*Owned	4.181	**0.027**
Brand*Sex*Owned	0.858	0.436

LOGO-startle reflex modulation	*F*	*p-value*
Brand	0.334	0.569
Brand*Sex	0.083	0.776
Brand*Owned	3.747	**0.038**
Brand*Sex*Owned	2.397	0.112

Significant values are indicated in Bold

Male Galaxy phone owners showed lower startle responses when presented with the Samsung logo compared to iPhone owners when presented to the Apple logo (see Fig. 2). However, in both owner groups there was no difference between the Apple and the Samsung logo. In females though, there was a clear tendency of iPhone owners to respond more positively to the Samsung logo and Galaxy owners to respond more positively to the Apple logo. Surprisingly, female other phone owners demonstrated more positive affective responses to both the Apple and the Samsung logos than did iPhone and Galaxy phone owners to their respective logos. See Fig. 2 for bar diagrams.

3.3 Pictures of Smartphones

Explicit rating performance shows that owners respond differently to Samsung and Apple smartphone depending on what smartphones they own. Among Apple owners, female and male participants both responded very similarly demonstrating higher preference for iPhones. In Samsung smartphone owners though, females showed higher preference for iPhones too, which was not the case in male Samsung owners.

Interestingly, an overall "version" effect occurred. This shows that the two different versions of each brand had varying effects on explicit ratings related to picture presentations of those phones.

Smartphone brand-valence rating	*F*	*p-value*
Brand	1.728	0.201
Brand*Sex	0.083	0.775
Brand*Owned	8.370	**0.002**
Brand*Sex*Owned	1.091	0.351
Version	10.037	**0.004**
Version*Sex	0.388	0.539
Version*Owned	1.742	0.196
Version*Sex*Owned	1.027	0.373
Brand*Version	0.566	0.459
Brand*Version*Sex	1.044	0.317
Brand*Version*Owned	1.342	0.280
Brand*Version*Sex*Owned	0.070	0.933

Significant values are indicated in Bold

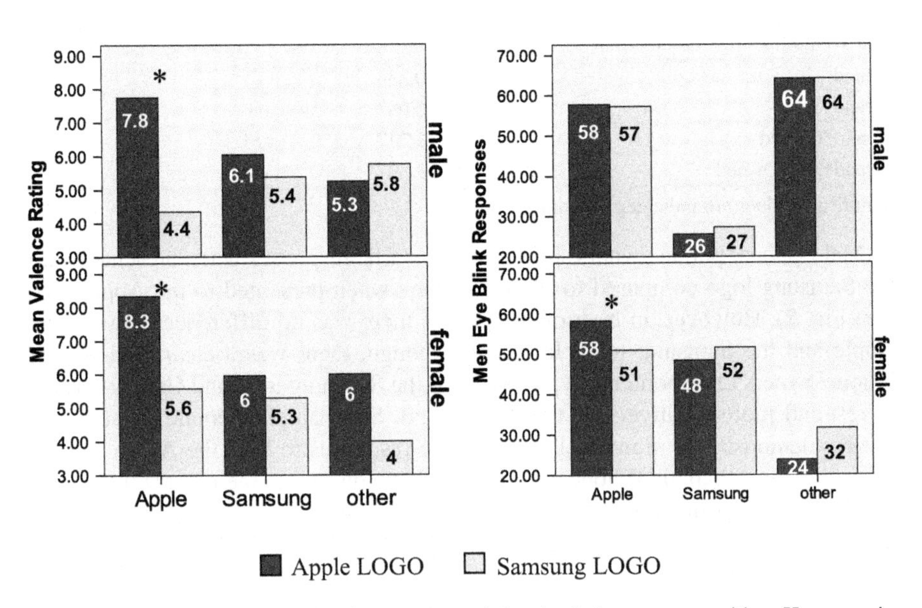

Fig. 2 Both female and male Apple owners rated the Apple logo more positive. However, in females startle responses show that the Samsung logo elicited more positive raw affective processing

Startle Reflex Modulation data show only a trend for a three-way interaction between brand, version and sex regardless of smartphone ownership. In particular, male participants responded more positively to the older Galaxy phone compared to the newer version (see Fig. 3).

Smartphone brand-startle reflex modulation	*F*	*p-value*
Brand	0.008	0.928
Brand*Sex	0.499	0.486
Brand*Owned	0.007	0.993
Brand*Sex*Owned	0.269	0.766
Version	1.490	0.234
Version*Sex	0.132	0.720
Version*Owned	0.112	0.895
Version*Sex*Owned	0.711	0.501
Brand*Version	0.000	0.999
Brand*Version*Sex	3.068	**0.092**
Brand*Version*Owned	0.182	0.834
Brand*Version*Sex*Owned	0.111	0.896

Significant value is indicated in Bold

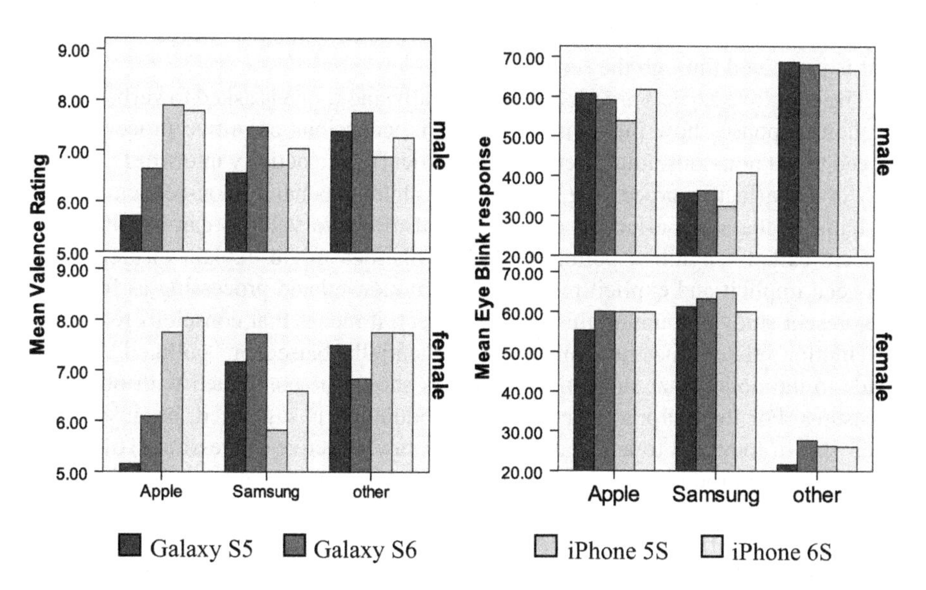

Fig. 3 Bar diagrams to show explicit responses and startle responses to visual presentations of Samsung and Apple phones

An overall outcome that is worth being mentioned is that male Samsung owners in general showed most positive affective processing (as in most reduced startle responses) regardless of brand, version and even experiment.

4 Discussion

This investigation also supports the idea that highly conscious and thought through explicit responses do not always match up with lower-level more implicit brain responses that are known to dominantly guide human behavior. This is particularly true for affective content (see [8, 9]). Pleasantness related to urban neighborhoods [10], different bottle shapes [11] and food [12], as well as emotion ownership [4] and emotion-related experimental design [7] have shown discrepancies between explicit versus implicit affective responses, while brand attitude for instance showed a match [13].

After all, it is mainly the discrepancies that brought to our attention that emotion-related information processing has different aspects [14, 15]. It has been highlighted that emotion is "not what you think it is" [2], that there is more to emotion than subjective feeling. As the main basic scientific progress we can assume and conclude that the human brain continuously processes affective information, which is coded via subcortical neural activity and which is primarily used to make affect-based decisions. In terms of evolution, it is a matter of pure survival to process affective information, because it's reflecting avoidance- and approach-related aspects of any stimulus. This is more important than cognitive aspects, which are reflective of the pure nature of any stimulus or in other words "what it is" that is perceived through the senses.

The cognitive aspects are processed cortically and if one is asked to verbalise via explicit responses how pleasant something is, conscious cognitive processes are forced to put non-conscious affective (subcortical) brain activity into words. On the way of raw affective processing, which guides human behavior non-consciously, to verbalising the same affective content initial affective valence can be altered, a process called "cognitive pollution" [13]. This idea might explain discrepancies between implicit and explicit responses to emotion-related processing as found in the present study. Assuming this idea is correct, it means that conscious responses to emotion-related questions are always potentially biased or "polluted", which leads to the notion that subjective responses should be completed with objective measures. For this purpose, startle reflex modulation is suggested, because it has been shown long ago to adequately quantify raw affective processing [16]. It has already been introduced to various different fields including distinct implications, clinical and industry-related [17–19]. The present study provides evidence that on a non-conscious affective level Apple and Samsung are not that differently processed, while conscious cognition deals with both brands in different ways.

Acknowledgements This study was conducted in collaboration with the Sigmund Freud University in Vienna.

References

1. Koller, M., Walla, P.: Towards alternative ways to measure attitudes related to consumption: introducing startle reflex modulation. J. Agric. Food Ind. Organ. **13**(1), 83–88 (2015)
2. Walla, P., Koller, M.: Emotion is not what you think it is: Startle Reflex Modulation (SRM) as a measure of affective processing in NeuroIs. In: NeuroIs Conference Proceedings, Springer. doi: 10.1007/978-3-319-18702-0_24 (2015)
3. Kunaharan, S., Walla, P.: Clinical neuroscience—towards a better understanding of non-conscious versus conscious processes involved in impulsive aggressive behaviours and pornography viewership. Psychology **5**, 1963–1966 (2014)
4. Walla, P., Rosser, L., Scharfenberger, J., Duregger, C., Bosshard, S.: Emotion ownership: different effects on explicit ratings and implicit responses. Psychology **3A**, 213–216 (2013). doi:10.4236/psych.2013.43A032
5. Walla, P., Koller, M., Meier, J.: Consumer neuroscience to inform consumers—physiological methods to identify attitude formation related to over-consumption and environmental damage. Frontiers in Human Neuroscience, 20 May 2014
6. Walla, P.: Non-conscious brain processes revealed by Magnetoencephalography (MEG). In: Pang, E. (ed.) Magnetoencephalography. ISBN: 978-953-307-255-5. InTech. doi: 10.5772/28211. http://www.intechopen.com/books/magnetoencephalography/non-conscious-brain-processes-revealed-by-magnetoencephalography-meg (2011)
7. Mavratzakis, A., Molloy, E., Walla, P.: Modulation of the startle reflex during brief and sustained exposure to emotional pictures. Psychology **4**, 389–395 (2013). doi:10.4236/psych.2013.44056
8. Walla, P., Mavratzakis, A., Bosshard, S.: Neuroimaging for the affective brain sciences, and its role in advancing consumer neuroscience. In: Fountas, K.N. (ed.) Novel Frontiers of Advanced Neuroimaging. ISBN: 978-953-51-0923-5. InTech. doi: 10.5772/51042 (2013)
9. Walla, P., Panksepp, J.: Neuroimaging helps to clarify brain affective processing without necessarily clarifying emotions. In: Fountas, K.N. (ed.) Novel Frontiers of Advanced Neuroimaging. ISBN: 978-953-51-0923-5. InTech. doi: 10.5772/51761 (2013)
10. Geiser, M., Walla, P.: Objective measures of emotion during virtual walks through urban neighbour-hoods. Appl. Sci. **1**(1), 1–11 (2011). doi:10.3390/app1010001
11. Grahl, A., Greiner, U., Walla, P.: Bottle shape elicits gender-specific emotion: a startle reflex modulation study. Psychology **7**, 548–554 (2012). doi:10.4236/psych.2012.37081
12. Walla, P., Richter, M., Färber, S., Leodolter, U., Bauer, H.: Food evoked changes in humans: startle response modulation and event-related potentials (ERPs). J. Psychophysiol. **24**(1), 25–32 (2010). doi:10.1027/0269-8803/a000003
13. Walla, P., Brenner, G., Koller, M.: Objective measures of emotion related to brand attitude: a new way to quantify emotion-related aspects relevant to marketing. PLoS One **6**(11), e26782 (2011). doi:10.1371/journal.pone.0026782
14. Montag, C., Walla, P.: Carpe Diem instead of losing your social mind: beyond digital addiction and why we all suffer from digital overuse. Cogent Psychol. **3**, 1157281 (2016)
15. Mavratzakis, A., Herbert, C., Walla, P.: Emotional facial expressions evoke faster orienting responses, but weaker emotional responses at neural and behavioural levels compared to scenes: a simultaneous EEG and facial EMG study. Neuroimage **124**, 931–946 (2016)
16. Cuthbert, B.N., Schupp, H.T., Bradley, M., McManis, M., Lang, P.J.: Probing affective pictures: attended startle and tone probes. Psychophysiology **35**, 344–347 (1998)
17. Lyons, G.S., Walla, P., Arthur-Kelly, M.: Toward improved ways of knowing children with profound multiple disabilities (PMD): Introducing startle reflex modulation. Dev. Neurorehabil. **16**(5), 340–344 (2013). doi:10.3109/17518423.2012.737039
18. Koller, M., Walla, P.: Measuring affective information processing in information systems and consumer research—introducing startle reflex modulation. In: ICIS Proceedings, Breakthrough Ideas, Full Paper in Conference Proceedings, Orlando. http://aisel.aisnet.org/icis2012/proceedings/BreakthroughIdeas/1/ (2012)

19. Nesbitt, K., Blackmore, K., Hookham, G., Kay-Lambkin, F., Walla, P.: Using the startle eye-blink to measure affect in players. In: Loh, C.S., Sheng, Y., Ifenthaler, D. (eds.) Serious Games Analytics, pp. 401–434. Springer, Switzerland (2015). doi:10.1007/978-3-319-05834-4_18

Neural Correlates of Technological Ambivalence: A Research Proposal

Hillol Bala, Elise Labonté-LeMoyne, and Pierre-Majorique Léger

Abstract Individuals' reactions to a new technology have been studied extensively in information systems and other literatures. We suggest that one of the possible dispositions to a new technology, ambivalence, has not been studied much in the information systems research. Ambivalence refers to a state of having both positive and negative orientations toward an object simultaneously. Our conjecture is that it is possible for an individual to have such a reaction toward a new technology. In addition to having both positive and negative orientations, ambivalence may also represent as cognition and/or emotion. We suggest that such a complex reaction and/or state can be better understood using a neurophysiological approach because this approach will help us disentangle various facets and temporal dynamics of ambivalence. In this proposed research, we seek to conduct an empirical study to understand the neural correlates of ambivalence toward a technology.

Keywords Ambivalence • Technology • EEG • Source localization

1 Introduction

Individuals' reactions to a new technology have been a topic of much interest in the information systems (IS) literature. Prior research has focused extensively on how individuals react to a new technology, such as beliefs (e.g., perceived usefulness, perceived ease of use, user involvement, user participation, social influence, and perceived enjoyment), attitudes (e.g., user satisfaction, attitude towards a technology), intentions (e.g., intention to use and continuance intention), and behaviors (e.g., system use, avoidance, exploration, exploitation, and resistance behaviors). Indeed, this stream of research has been considered one of the most prolific and mature streams of research in IS [1, 2]. Notwithstanding our extensive

H. Bala
Indiana University, Bloomington, IN, USA
e-mail: hbala@indiana.edu

E. Labonté-LeMoyne (✉) • P.-M. Léger
Tech3Lab, HEC Montreal, Montreal, QC, Canada
e-mail: elise.labonte-lemoyne@hec.ca; pml@hec.ca

F.D. Davis et al. (eds.), *Information Systems and Neuroscience*, Lecture Notes in Information Systems and Organisation 16, DOI 10.1007/978-3-319-41402-7_11

understanding of individuals' reactions to a new technology, technology implementations still fail in organizations and people issues are still considered a major concern in technology implementation projects. For example, a recent industry report has touted people-related issues, such as executive sponsorship, emotional maturity, and user involvement, as the top three success factors in IS implementation projects [3].

Prior IS research has primarily focused on either a positive (e.g., opportunity, acceptance, perceived usefulness, perceived enjoyment) or a negative (e.g., threat, resistance, perceived anxiety, technostress) disposition toward a new technology. Although there are few exceptions that have focused on mixed reactions (e.g., [4]), we suggest that such a dichotomous view (i.e., either positive or negative) is perhaps narrow and does not reflect the fact that many individuals may have an ambivalent view towards a new technology (i.e., both positive and negative dispositions at the same time). Indeed, research in social psychology and organizational behavior has suggested that the "experience of simultaneously positive and negative orientations toward a person, goal, task, idea, and such appears to be quite common…but it is poorly understood" [5]. *Ambivalence* is defined as simultaneously oppositional positive and negative orientations toward an object [5, 6]. In the context of a technology in the workplace, it is possible that an individual may think it will make his or her work more effective—i.e., a positive orientation. At the same time, the same individual may think the technology will be hard to master and it will change his or her routines that have made him or her successful in the first place—i.e., a negative orientation. Such an ambivalent view is possible for both work-related (i.e., utilitarian) and non-work-related (i.e., personal and/or hedonic) technologies. We suggest that in today's environment in which individuals are exposed to a wide variety of ubiquitous and pervasive technologies in their personal and professional lives, it is likely that ambivalence will be a prevalent individual disposition towards a technology.

Prior research has paid limited attention to the notion of ambivalence or mixed reactions toward technologies. In a recent article, Stein et al. [4] found that individuals developed ambivalent emotions that led to task and technology adaptation behaviors. Beaudry and Pinsonnault [7] studied technology-related coping behaviors and suggested that individuals may develop perceptions of opportunity and threat. Although they alluded to the notion of simultaneous presence of opportunity and threat perceptions, they did not explicitly theorize the notion of ambivalence in their study. More recently, they did study the effects of both positive (i.e., happiness and excitement) and negative (i.e., anger and anxiety) emotions on coping behaviors and technology use [8]. Recently, Bala and Venkatesh [9] also studied opportunity and threat reactions and subsequent adaptation behaviors. However, these studies primarily focused on behavioral aspects of technology adaptation, examined positive and negative reactions and/or emotions separately, and did not explicitly theorize and test the notion of ambivalence toward a technology, i.e., simultaneous presence of positive and negative orientations toward a technology.

While it is possible to study ambivalence using the traditional behavioral approach (e.g., measuring ambivalence using survey items), recent research in social psychology has suggested that neurophysiological measurement of ambivalence will provide a better understanding of cognitive and emotional state involved in ambivalence [10]. By definition, ambivalence is the synchronous presence of two or more emotional and cognitive states. Such complex cognitive and emotional states can be better captured and understood through neurophysiological measures than traditional questionnaire/survey based approaches. Through the use of electroencephalography, we will be able to tease apart these states and their sequence (and/or simultaneity) even when they are subconscious. Our research questions are as follows:

1. *What are the neural correlates of ambivalence?*
2. *How is ambivalence towards technology neurophysiologically different from only a positive or negative state?*
3. *Can we tease apart the sequence in which the positive and negative states occur when ambivalence is present?*

In order to find answers to these research questions, we propose to conduct an empirical study in which individuals will be exposed to several work and non-work-related technologies. One of our goals is to develop and present a methodology for a neurophysiological measure of ambivalence. We plan to experimentally manipulate positive, negative, and ambivalent orientations toward these technologies. We expect to invoke these orientations and measure how different parts of the human brain get activated by these orientations. We expect to find that ambivalence is indeed a distinct neurological reaction and that individuals develop ambivalent attitudes toward a technology. We also expect to unearth how ambivalence influences individuals' intentions to use a technology. Our research is expected to make substantial contributions by offering insights on ambivalence toward technology, an individual reaction, that has not been studied much in the IS literature.

2 Hypotheses

Given that ambivalence is an experience with both positive and negative orientations, we expect that it will activate different regions of brain. At a minimum, we expect that ambivalence will activate more brain areas that either positive or negative orientations can activate individually. In addition to having both positive and negative orientations toward a technology, ambivalence may also include cognition ("I think about technology x) and emotion ("I feel about technology x") simultaneously [5]. Therefore, we expect that ambivalence will activate not only different brain regions, but also more brain areas compared to positive and negative orientations toward technology.

H1a Ambivalence toward a technology elicits multiple different but simultaneous sources of brain activity in individuals (including activity that is not present when in presence of either positive or negative orientations toward a technology).

H1b Ambivalence toward technology elicits more cerebral activity in individuals than either positive or negative orientations toward technology.

We suggest that ambivalence will have a negative influence on intention to use a new technology. Prior research has found that ambivalence typically leads to negative outcomes because it is difficult for individuals to reconcile the clash between positive and negative orientations and between cognitions and emotions [5, 10]. Individuals are likely to feel tensed and confused when they develop ambivalence toward a technology. Therefore, we expect that ambivalence will have a negative influence on intention to use a new technology.

H2 Ambivalence toward technology will have a negative influence on intention to use a new technology.

3 Proposed Methodology

The proposed methodology builds upon the traditional neuroscientific method of eliciting event-related potentials. Event-related potential research is meant to determine the sequence of the brain's electrical activity in response to a presented stimulus. By presenting stimuli of a certain type a large number of times and averaging together the brain responses immediately following each stimulus, the background brain activity is averaged out and we are left with the brain response that is specific to that type of stimulus. We can also employ source localization algorithms to triangulate the location of the response in the brain and thus better understand the underlying cognitive processes [11–13].

3.1 Participants

Forty participants will be recruited through our institution's recruitment panel. Usual demographics for this panel consist of university students aged between 18 and 35 years old and equally distributed between genders. The study will be presented to the institutional review board before it begins.

3.2 Experimental Procedure

The first task is based upon an ambivalence study by Nohlen et al. [10]. Participants will first be presented with the name of a technology accompanied by two words that describe characteristics of this technology. The participant will then be asked to answer a yes/no question about the technology, specifically, "Would you adopt this technology?". They will answer with two buttons, one for yes, and one for no. Positive (for example: it will make you more effective) and negative (for example: it will be hard to master) characteristics will be combined as either congruent pairs (positive–positive or negative–negative) or incongruent pairs (positive–negative). Incongruent pairs are designed to induce ambivalence.

Each trial will start with a fixation cross (3000 ms) followed by the presentation of the name of the technology accompanied by its two characteristics until the participant answers. Three blocks of 60 trials each (approximately 10 min each) will be presented with a 5 min break in between each block to adjust equipment and allow the participant to rest. Each block will have 20 congruent positive trials, 20 congruent negative trials and 20 incongruent trials. The order of trials will be randomized and the buttons will be counterbalanced.

No specific evoked potential markers of ambivalence have been identified to this day, however we can use two brain responses that are sensitive to decision making and emotional valence given the right stimulus. In cognitive psychology literature, it has been demonstrated that more difficult choices elicit smaller P3 event-related potentials [14]. We can thus pose the hypothesis that incongruent trials will induce more ambivalence and that ambivalence would be directly correlated with the amplitude of the P3 event-related potential. We can also pose the hypothesis that congruent trials will elicit frontal alpha asymmetry which has been linked to emotional valence [15] while incongruent trials, through the eliciting of both positive and negative emotions, might present less asymmetry.

3.3 EEG Acquisition and Analysis

The electroencephalography (EEG) will be recorded with a 128 channel Brain Products Actichamp active electrode system (Brain Products, Munich, Germany). The data will be acquired at 1000 Hz with no acquisition filters, and the built-in FCz reference. It will then be offline preprocessed by filtering, rereferencing, artifact removing, segmenting and averaging for event-related potentials or frequency decomposing with wavelets for frontal alpha asymmetry.

The event-related potentials and the frontal alpha asymmetry will be compared between congruent and incongruent trials to answer H1. The correlation between the amplitude of the event-related potentials and the participant's answers as well as their reaction times to answer will be calculated to answer H2.

Finally, source localization will be employed to locate the source of the cerebral activity measured. The deartefacted EEG signal will be decomposed through independent component analysis (ICA) and we will apply clustering analysis to the components. Following that, the dipole fitting (DIPFIT) tool from EEGlab [16–19] will be used to identify the location in the brain of the source dipole for each cluster. This will inform us of the areas of the brain that are differently active for congruent and incongruent pairs of stimuli. With source localization, we could find that these regions present separate, but simultaneous, coactivated measures of positive and negative emotions, similar to those observed in the solely positive/negative conditions, or they could present a new and distinct marker of ambivalence.

4 Discussion and Expected Contributions

Ambivalence has been studied extensively in social psychology and organizational behavior literatures. It is a prevalent individual reaction that has been largely overlooked in the IS literature. We suggest that it is possible for individuals to develop an ambivalent view toward a technology and such a view will have an effect on their adoption and use of the technology. Notwithstanding the maturity of research on individuals' reactions toward technology and their adoption and use behaviors, we remain limited in our understanding of these important phenomena if we do not theorize and test why and how individuals form ambivalent views toward a technology and the impact of technological ambivalence.

Our goal is to develop a deep understanding of technological ambivalence using a NeuroIS perspective. Given that ambivalence is a complex individual disposition (i.e., simultaneous presence of positive and negative orientations), we suggest that a neurophysiological examination of ambivalence toward a technology will provide insights on how technological ambivalence is different from a purely positive or negative orientation toward a technology. Our study is expected to provide insights on how ambivalence activates human brain areas and this will help us understand the neurophysiological sequencing involved in the development of ambivalence. Further, our study is expected to provide insights on how ambivalence is different from positive and negative orientations that have been studied extensively in the IS literature and how it affects individuals' intention to use a new technology. We expect to find a strong presence of ambivalence as a salient reaction to a new technology and a negative association between ambivalence and the intention to use a new technology. This suggests that ambivalence is an important individual reaction toward a technology that has both theoretical and practical significance. We expect that our work will open new avenues of research on individuals' reactions to a technology.

References

1. Venkatesh, V.: Where to go from here? Thoughts on future directions for research on individual-level technology adoption with a focus on decision making. Decis. Sci. **37**, 497–518 (2006)
2. Venkatesh, V., Morris, M.G., Davis, G.B., Davis, F.D.: User acceptance of information technology: toward a unified view. MIS Q. **27**, 425–478 (2003)
3. Standish Group.: 2015 Chaos Report. (2015)
4. Stein, M.-K., Newell, S., Wagner, E.L., Galliers, R.D.: Coping with information technology: mixed emotions, vacillation, and nonconforming use patterns. MIS Q. **39**, 367–392 (2015)
5. Ashforth, B., Rogers, K., Pratt, M., Pradies, C.: Ambivalence in organizations: a multilevel approach. Organ. Sci. **25**, 1453–1478 (2014)
6. Oreg, S., Sverdlik, N.: Ambivalence toward imposed change: the conflict between dispositional resistance to change and the orientation toward the change agent. J. Appl. Psychol. **96**, 337–349 (2011)

7. Pinsonneault, A., Beaudry, A.: Understanding user responses to information technology: a coping model of user adaptation. MIS Q. **29**, 493–524 (2005)
8. Beaudry, A.: The other side of acceptance: studying the direct and indirect effects of emotions on information technology use. MIS Q. **34**, 689–710 (2010)
9. Bala, H., Venkatesh, V.: Adaptation to information technology: a holistic nomological network from implementation to job outcomes. Manag. Sci. **62**, 156–179 (2015)
10. Nohlen, H.U., van Harreveld, F., Rotteveel, M., Barends, A.J., Larsen, J.T.: Affective responses to ambivalence are context-dependent: a facial EMG study on the role of inconsistency and evaluative context in shaping affective responses to ambivalence. J. Exp. Soc. Psychol. **65**, 42–51 (2016)
11. Müller-Putz, G.R., Riedl, R., Wriessnegger, S.: Electroencephalography (EEG) as a research tool in the information systems discipline: foundations, measurement, and applications. Commun. Assoc. Inf. Syst. **37**. http://aisel.aisnet.org/cais/vol37/iss1/46 (2015).
12. Riedl, R., Léger, P.-M.: Fundamentals of NeuroIS. Springer, Berlin (2016)
13. Luck, S.J., Kappenman, E.S. (eds.): The Oxford Handbook of Event-Related Potential Components. Oxford University Press, New York (2011)
14. Cutmore, T.R.H., Muckert, T.D.: Event-related potentials can reveal differences between two decision-making groups. Biol. Psychol. **47**, 159–179 (1998)
15. Harmon-Jones, E., Gable, P.A., Peterson, C.K.: The role of asymmetric frontal cortical activity in emotion-related phenomena: a review and update. Biol. Psychol. **84**, 451–462 (2010)
16. Delorme, A., Palmer, J., Onton, J., Oostenveld, R., Makeig, S.: Independent EEG sources are dipolar. PLoS One **7**(2), e30135 (2012)
17. Makeig, S., Debener, S., Onton, J., Delorme, A.: Mining event-related brain dynamics. Trends Cogn. Sci. **8**, 204–210 (2004)
18. Delorme, A., Mullen, T., Kothe, C., Akalin Acar, Z., Bigdely-Shamlo, N., Vankov, A., Makeig, S.: EEGLAB, SIFT, NFT, BCILAB, and ERICA: new tools for advanced EEG processing. Comput. Intell. Neurosci. **2011** (2011)
19. Delorme, A., Makeig, S.: EEGLAB: an open source toolbox for analysis of single-trial EEG dynamics including independent component analysis. J. Neurosci. Methods **134**, 9–21 (2004)

Neuroscience Foundations for Human Decision Making in Information Security: A General Framework and Experiment Design

Bin Mai, Thomas Parsons, Victor Prybutok, and Kamesh Namuduri

Abstract In this paper, we propose a conceptual model for information security (InfoSec) decision making that is based on cognitive science and neuroscience, and present a comprehensive framework of InfoSec decision making process. In addition, we illustrate a specific experiment design that could be used to investigate how specific automatic affective processes would impact specific components of the InfoSec decision making process.

Keywords Information security • Decision making • Neuroscience • NeuroIS • Conceptual framework • Affective executive control • Cognitive executive control • Experiment design

1 Introduction

The main focus of this paper is to present a basic framework for applying neuroscience concepts, theories, and tools to studying human decision making in information security situations. Information security (InfoSec) has been a significant issue in the society, exacerbated by the current "big data" environment in today's global society [1]. By the very nature of information security, this field has been an intersection among engineering technology, economic incentive design, and human behavior [2]. Among the various aspects of InfoSec issues, human decision making during real InfoSec situations remains one of the most critical and under-investigated issues in InfoSec [3, 4]. After all, it is human decision making that ultimately determines the outcomes of InfoSec planning and implementation. By understanding the details how individuals make such decisions and how various factors could influence these decision making processes, we would be able to design better information systems artifacts, develop better policies and regulations, provide better training/education, construct a better culture and environment, all from

B. Mai (✉) • T. Parsons • V. Prybutok • K. Namuduri
University of North Texas, Denton, TX, USA
e-mail: bin.mai@unt.edu; thomas.parson@unt.edu; victor.prybutok@unt.edu; Kamesh.namuduri@unt.edu

F.D. Davis et al. (eds.), *Information Systems and Neuroscience*, Lecture Notes in Information Systems and Organisation 16, DOI 10.1007/978-3-319-41402-7_12

the perspectives of facilitating the individuals to make better InfoSec decisions and thus achieve more desirable InfoSec outcomes.

Traditionally, IS researchers interested in the determinants of behavior have emphasized controlled (conscious) processing of perceptions and intentions over automatic (unconscious) processing of information. Quite often, researchers have studied InfoSec's controlled decision making processes by analyzing data collected through mostly surveys and interviews, constructing empirical models for decision making process, and developing analytical and quantitative models to derive optimal decision strategies, all based on the standard assumption of the decision makers being rational agents [5–7]. While this stream of research has yielded valuable insights towards InfoSec decision making, there exist significant gaps which are not well explained by the current methodologies [8].

Recently IS research has begun to emphasize the role that habit plays in the post-adoption stage of usage, in which the behavior turn out to be more automatic and may be performed with a reduced amount of controlled (conscious) processing. Researchers have also studied the role of habit in InfoSec contexts [9–11]. However, one concern for current approaches to studying InfoSec decision making is that the use of self-report methods are insufficient for assessing user mental state given the profound limitations of survey methods for measuring relatively automatic (unconscious) heuristics and biases as they are less accessible to introspection [12]. Moreover, in search for more objective and comprehensive approaches to studying InfoSec decision making, researchers are increasingly embracing advances in neuroscience theories, techniques, and tools, such as psychophysiological metrics and neuroimaging techniques, because they offer promise in alleviating the limitations found in subjective self-reports through more direct and objective measurement of automatic processes [13]. Still, what are lacking in the InfoSec literature include a general framework to facilitate the investigation of InfoSec decision making with these new approaches, and some concrete examples for implementing this framework.

In this paper, following the NeuroIS Research Framework proposed by [14], we formulate our research question as: how can we systematically incorporate relevant neuroscience theories, techniques, and tools into the study of InfoSec decision making to make it more objective, accurate, and realistic? In addressing this question, we make three contributions to the literature of InfoSec decision making. First we propose a general framework for InfoSec decision making that is based on cognitive science and neuroscience. Second, in our propose framework, we explicitly emphasize the different roles played by controlled cognitive processes and automatic affective processes. And last but not least, we describe a specific experiment design that could be used to investigate how specific automatic affective processes would impact specific components of the InfoSec decision making process.

2 A General Framework for Neuro-Scientific InfoSec Decision Making

InfoSec decision making is one specific type of general decision making. However, there are two components that are particularly significant in the InfoSec decision making context: risk and uncertainty [15]. Within InfoSec context, risk often refers to the likelihood that information assets are insufficiently protected against certain types of damages or loss [16]; and uncertainly usually refers to one form of manifestation of information asymmetry [17]. Traditionally, a widely accepted model for decision making has been the Observe–Orient–Decide–Act (OODA) Loop proposed by [18]. It is a model originally design for understanding decision making processes adopted by military commanders to enhance command and control. However, as [19] pointed out, the OODA model "has a number of problems as a framework of human decision making" including the failure to account for the "necessary dependence of perception on preexisting knowledge and concepts" during decision making. To address the deficiencies of the OODA paradigm, [19] proposed a cognitive science based framework of decision making, the CECA (Critique, Explore, Compare, Adapt) framework to model individual decision making. Although both OODA and its extension CECA originated within a military context, their acceptance and application have since gone beyond the military decision making situations, and have become an important decision model in a wide range of decision making scenarios including emergency response [20], system safety development [21], and significantly, InfoSec [3], in which it is pointed out that OODA and CECA "could be a usable framework for the security manager's decision process" (p. 60). In addition, the CECA framework specifically addresses the issues of risk and uncertainty. Therefore, in this paper we integrate this CECA paradigm with the OODA framework elements to model InfoSec decision making processes. The overall framework for the InfoSec decision making can thus be illustrated by the following figure (Fig. 1):

More specifically, the InfoSec decision making process would include the following procedures:

1. Situation analysis—(Observe/Orient)
   (a) Recognizing whether this is an infosec situation (formulate questions/identify info needs, define security mission/develop security goals)
2. Risk analysis (Explore)
   (a) Threat analysis (what are the threats; the probability of the threats being realized; the damage if the threats are realized)
   (b) Counter measure analysis (what are the counter measures; the effectiveness of the counter measures; the costs for implementing the counter measures)
3. Trade-off analysis (Compare/Decide)
   (a) Alternatives analysis (what are the available alternatives of actions to take)

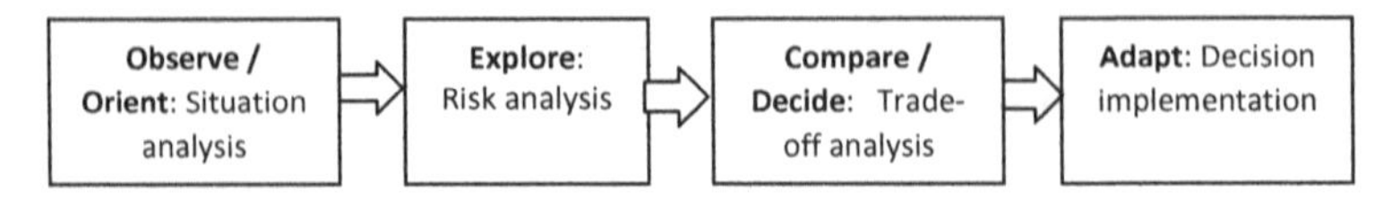

Fig. 1 The framework for InfoSec decision making

   (b) Mental algebra (compute and compare the overall expected payoffs/costs of each alternative)
   (c) Decide on which alternative to take

4. Implement the decision (Adapt)

   (a) Actually implement the decision once it is made.

The above framework places significant emphasis on the role of information collection in each of the steps during decision making process, and thus focus more on the factors of controlled cognitive processes and their influence on the decision making process. However, as pointed out in previous section, the uncontrolled and automatic processes also play a significant role in influencing each of the steps of the decision making process. We need to extend the above framework to explicitly incorporate such factors to complement the existing controlled processes and provide a more comprehensive picture of InfoSec decision making.

One field that holds great promise in contributing to the above goal of InfoSec decision making framework is neuroscience, which is the science that explains how human brain works. A component missing from most InfoSec scenarios research is the reality that controlled decision making may include affective recognition and evaluation of stimuli. Affective stimuli are particularly potent distracters that can reallocate processing resources and impair attentional performance [22]. In addition to the cognitive appraisal of a stimulus, affective reactions are characterized by psychophysiological changes (e.g., alterations in skin conductance and heart rate; as well as behavioral approach or avoidance) and involve a number of subcomponents occurring in frontal subcortical circuits [23–25]. According to models of neurovisceral integration, autonomic, attentional, and affective systems are simultaneously engaged in the support of self-regulation [26, 27].

Based on the above neuroscience findings, we expand the above general InfoSec decision making process model via the inclusion of a dual-process approach to modeling decision making, in which both automatic affective and cognitive processes perform essential roles in controlled decision making: (1) automatic processing is reflexive, rapid, and led by heuristics and biases. They operate automatically, intuitively, involuntarily, and effortlessly; and (2) controlled cognitive processing refer to the people's reflective, logical, and rational decision making.

Therefore, our proposed general framework for neuro-scientific InfoSec decision making can be illustrated by the following figure (Fig. 2):

With this framework, we have established the foundation for many possible future research projects of InfoSec decsion making: (1) investigate and determine the neural

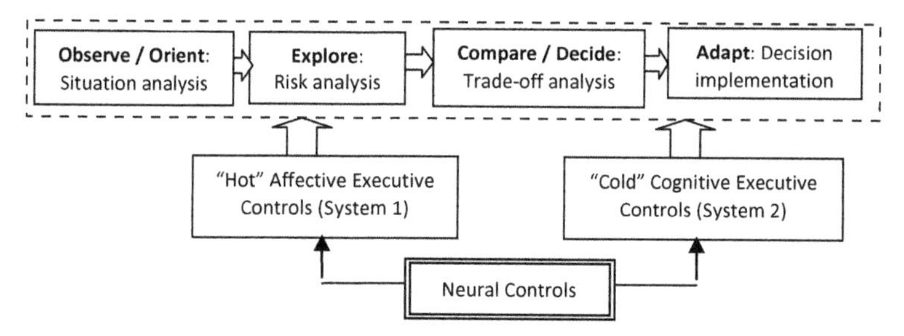

Fig. 2 A general framework for neuro-scientific InfoSec decision making

correlates (brain areas) of automatic and controlled decision making during InfoSec situations; (2) investigate and develop comprehensive models for automatic, as well as controlled, human decision making during InfoSec situations, aiming to incorporate/map the neural correlates of decision making processes; (3) identify possible automatic affective and cognitive biases and associated neural processes during the controlled infoSec decision making; and (4) design and develop mechanisms, based on theoretical foundations for decision making, to influence human decision making during information security situations, in order to overcome identified affective and cognitive biases and achieve optimal information security outcomes.

3 A Research Experiment Design

In this proposed project, we aim to establish an approach to cognitive (task engagement) and affective (arousal) state estimation for cyber threat detection. Based on our proposed model, this is a topic of the impacts of System 1 and 2 factors on the observe/orient process of InfoSec decision making. The psychophysiological signals from users could be logged as participants experience various stimulus modalities aimed at assessing automatic and controlled cognitive and affective processing. Psychophysiological metrics and event-related potentials (ERP) have been used to infer various aspects of an individual such as personality, memory and preferences. More specifically, the following methodologies can be adopted:

- *Psychophysiological Metrics for Establishing Indices of Arousal (System 1)*: Psychophysiological techniques (e.g., cardiovascular reactivity, skin conductance response, respiration, EEG source models of connectivity) allow for enhanced understanding of the ways in which emotion and cognition interact [28, 29]. In addition to autonomic measures and EEG spectral power and waveforms, interactions between pairs of EEG oscillations—such as phase synchronization and coherence—have also been implicated in affective states of hedonic arousal [30].

- *Psychophysiological Metrics for Establishing Indices of Cognitive Control (System 2)*: Using psycho-physiological metrics to measure cognitive processing has been widely used. Results suggest that autonomic and EEG engagement reflects information gathering, visual processing, and attention allocation [31].
- *Data Analytics*: The psychophysiological data could then be filtered to get separate frequency bands to train cognitive-affective classifiers with classification techniques such as support vector machines, naïve Bayes, and k nearest neighbors [32].

A "Psychophysiological Baseline" would be established for each participant. To establish each participant's "Psychophysiological Baseline" (*minimum* brain wave and autonomic activity) participants would be told "to relax and try not to think about anything" as they stare at a blank screen for 2:00 min. Next, the participant would take part in the following cognitive workload and arousal tasks to establish *maximal* physiological responses: (1) a "Two Picture Cognitive Task" would be used to establish a cognitive workload signature for the user [33]; (2) the International Affective Picture System (IAPS) would be used to establish the affective load signature of the user; and (3) the cyber threat scenario. Task presentation order would be counter-balanced across participants.

To assess psychophysiological reactivity to the cyber threat scenario, we would compute reactivity scores for cognitive workload and affective load by (1) subtracting "Psychophysiological Baseline" values from the cyber threat scenario; and (2) comparing Cognitive and Arousal indices to values from the insider threat scenario.

Due to length limit of the paper. We omit the description of our data analysis process.

4 Conclusion

In this paper, we proposed a general framework for neuro-scientific InfoSec decision making. The framework is based on cognitive science and incorporates two important categories of neural controls that would impact the InfoSec decision making processes: automatic affective executive controls and controlled cognitive executive controls. We illustrate how this proposed framework can be the foundation for the development of a comprehensive model of neuro-scientific InfoSec decision making, and discuss some specific research questions that can be addressed through our framework. In addition, we provide a detailed description of a scientific research experiment project that is based on our proposed conceptual framework.

The immediate next step would be to embellish details of our basic framework, and accordingly identify specific automatic and controlled executive controls significant in InfoSec decision making behaviors. We would then hypothesize the functioning mechanisms of those controls in terms of how they impact the InfoSec decision making, and conduct lab experiment to empirically investigate those mechanisms.

References

1. Wang, H., Jiang, X., Kambourakis, G.: Special issue on security, privacy and trust in network-based big data. Inf. Sci. **318**, 48–50 (2015)
2. Chatterjee, S., Sarker, S., Valacich, J.S.: The behavioral roots of information systems security: exploring key factors related to unethical IT use. J. Manag. Inf. Syst. **31**, 49–87 (2015)
3. Pettigrew III, J.A., Ryan, J.J.: Making successful security decisions: a qualitative evaluation. IEEE Secur. Priv. **10**(1), 60–68 (2012)
4. Schneier, B.: The psychology of security. In: Vaudenay, S. (ed.) Progress in Cryptology–AFRICACRYPT 2008, pp. 50–79. Springer, Berlin (2008)
5. Arora, A., Nandkumar, A., Telang, R.: Does information security attack frequency increase with vulnerability disclosure? An empirical analysis. Inf. Syst. Front. **8**, 350–362 (2006)
6. Fang, F., Parameswaran, M., Zhao, X., Whinston, A.B.: An economic mechanism to manage operational security risks for inter-organizational information systems. Inf. Syst. Front. **16**, 399–416 (2014)
7. Hsu, J.S.-C., Shih, S.-P., Hung, Y.W., Lowry, P.B.: The role of extra-role behaviors and social controls in information security policy effectiveness. Inf. Syst. Res. **26**, 282–300 (2015)
8. Acquisti, A., Grossklags, J.: Privacy and rationality in individual decision making. IEEE Secur. Priv. **3**(1), 26–33 (2005)
9. Ng, B.-Y., Kankanhalli, A., Xu, Y.C.: Studying users' computer security behavior: a health belief perspective. Decis. Support Syst. **46**, 815–825 (2009)
10. Pahnila, S., Siponen, M., Mahmood, A.: Employees' behavior towards IS security policy compliance. In: 40th Annual Hawaii International Conference on System Sciences. HICSS 2007, pp. 156b–156b. IEEE (2007)
11. Vance, A., Siponen, M., Pahnila, S.: Motivating IS security compliance: insights from habit and protection motivation theory. Inf. Manag. **49**, 190–198 (2012)
12. Woodside, A.G.: Overcoming the illusion of will and self-fabrication: going beyond naïve subjective personal introspection to an unconscious/conscious theory of behavior explanation. Psychol. Mark. **23**, 257–272 (2006)
13. Pavlou, P., Davis, F., Dimoka, A.: Neuro IS: the potential of cognitive neuroscience for information systems research. In: ICIS 2007 Proceedings 122 (2007)
14. vom Brocke, J., Liang, T.-P.: Guidelines for neuroscience studies in information systems research. J. Manag. Inf. Syst. **30**, 211–234 (2014)
15. West, R.: The psychology of security. Commun. ACM **51**, 34–40 (2008)
16. Straub, D.W., Welke, R.J.: Coping with systems risk: security planning models for management decision making. Mis Q. **22**(4), 441–469 (1998)
17. Grossklags, J., Christin, N., Chuang, J.: Secure or insure?: a game-theoretic analysis of information security games. In: Proceedings of the 17th International Conference on World Wide Web, pp. 209–218. ACM (2008)
18. Boyd John, R.: A discourse on winning and losing. Air University document MU43947, briefing 1 (1987)
19. Bryant, D.J.: Rethinking OODA: toward a modern cognitive framework of command decision making. Mil. Psychol. **18**, 183 (2006)
20. Romanowski, C., Raj, R., Schneider, J., Mishra, S., Shivshankar, V., Ayengar, S., Cueva, F.: Regional response to large-scale emergency events: building on historical data. Int. J. Crit. Infrastruct. Prot. **11**, 12–21 (2015)

21. Trotter, M.J., Salmon, P.M., Lenne, M.G.: Impromaps: applying Rasmussen's risk management framework to improvisation incidents. Saf. Sci. **64**, 60–70 (2014)
22. Dolcos, F., McCarthy, G.: Brain systems mediating cognitive interference by emotional distraction. J. Neurosci. **26**, 2072–2079 (2006)
23. Bonelli, R.M., Cummings, J.L.: Frontal-subcortical circuitry and behavior. Dialogues Clin. Neurosci. **9**, 141 (2007)
24. Pessoa, L.: How do emotion and motivation direct executive control? Trends Cogn. Sci. **13**, 160–166 (2009)
25. Ray, R.D., Zald, D.H.: Anatomical insights into the interaction of emotion and cognition in the prefrontal cortex. Neurosci. Biobehav. Rev. **36**, 479–501 (2012)
26. Critchley, H.D.: Neural mechanisms of autonomic, affective, and cognitive integration. J. Comp. Neurol. **493**, 154–166 (2005)
27. Thayer, J.F., Lane, R.D.: A model of neurovisceral integration in emotion regulation and dysregulation. J. Affect. Disord. **61**, 201–216 (2000)
28. Okon-Singer, H., Hendler, T., Pessoa, L., Shackman, A.J.: The neurobiology of emotion-cognition interactions: fundamental questions and strategies for future research. Front. Hum. Neurosci. **9**, 58 (2015)
29. Wu, D., Courtney, C.G., Lance, B.J., Narayanan, S.S., Dawson, M.E., Oie, K.S., Parsons, T.D.: Optimal arousal identification and classification for affective computing using physiological signals: virtual reality Stroop task. IEEE Trans. Affect. Comput. **1**, 109–118 (2010)
30. Wyczesany, M., Grzybowski, S.J., Barry, R.J., Kaiser, J., Coenen, A.M., Potoczek, A.: Covariation of EEG synchronization and emotional state as modified by anxiolytics. J. Clin. Neurophysiol. **28**, 289–296 (2011)
31. Parsons, T.D., Courtney, C.G., Dawson, M.E.: Virtual reality Stroop task for assessment of supervisory attentional processing. J. Clin. Exp. Neuropsychol. **35**, 812–826 (2013)
32. Wu, D., Lance, B.J., Parsons, T.D.: Collaborative filtering for brain-computer interaction using transfer learning and active class selection. PLoS One **8**, e56624 (2013)
33. McMahan, T., Parberry, I., Parsons, T.D.: Modality specific assessment of video game player's experience using the Emotiv. Entertain. Comput. **7**, 1–6 (2015)

Combining Vicarious and Enactive Training in IS: Does Order Matter?

Félix G. Lafontaine, Pierre-Majorique Léger, Élise Labonté-LeMoyne, Patrick Charland, and Paul Cronan

Abstract The objective of the article is to provide empirical support for curriculum development to instructors using enactive learning in IS. Specifically, we are interested in understanding which instructional design, combining enactive and vicarious learning, leads to the most effective learning achievement and development of self-efficacy. Specifically, we compare two different training sequences to determine which is the best combination of the two instructional designs (vicarious/enactive) to train people in using business dashboards efficiently. In a controlled lab environment, we collected (1) behavioral data (performance, software interactions) (2) oculometric data and (3) self-assessed self-efficacy data to assess the learning processes and strategies. Our results show that providing the vicarious training first when using a combination of enactive and vicarious learning leads to a higher self-efficacy increase. It also has a significant impact on the attentional efficiency of students using dashboards in a business setting.

Keywords End user training • ERPsim • Business intelligence • Eye tracking

1 Introduction

End-user training is recognized to be a key factor in success of information system implementation [1, 2]. Training end-users for the business intelligence and analytics technologies will be especially important since Gartner predicts that by 2017, most business users will have access to technologies that will enable them to

F.G. Lafontaine (✉) • P.-M. Léger • É. Labonté-LeMoyne
HEC Montréal, Montréal, QC, Canada
e-mail: felix.gaudet-lafontaine@hec.ca; pierre-majorique.leger@hec.ca; elise.labonte-lemoyne@hec.ca

P. Charland
Université du Québec à Montréal (UQAM), Montréal, QC, Canada
e-mail: charland.patrick@uqam.ca

P. Cronan
University of Arkansas, Fayetteville, AR, USA
e-mail: cronan@uark.edu

F.D. Davis et al. (eds.), *Information Systems and Neuroscience*, Lecture Notes in Information Systems and Organisation 16, DOI 10.1007/978-3-319-41402-7_13

prepare and analyze data [3]. Research suggest that enactive learning is a very effective way to engage new users with a system [4]. Enactive learning implies that the learning is a consequence of one's interaction with and feedback from the environment [5]. However, research suggests that a combination of enactive learning and vicarious learning leads to greater learning outcomes compared to vicarious training alone [6, 7]. Vicarious learning suggests that the trainees learn by reflecting on their observation of someone performing a targeted behavior [8] which means that one can learn by observing the actions of another person and the associated consequences [6].

If the best strategy is to combine enactive and vicarious training as suggested by research [6, 7], it is important to determine the optimal order one should use these instructional designs in a IS training curriculum. Is it better to start with vicarious training activities followed by an enactive experience, or vice versa? We propose to answer this question by investigating the attentional efficiency of the participant. We conducted an eye tracking study to study in at which we controlled the order in which the participants received the vicarious and the enactive training.

2 Literature Review and Hypothesis

The eye movement research is based on the eye-mind assumption proposed by Just and Carpenter [9], according to which the attention is closely linked to the direction of the gaze. This assumption is valid as long as the visual environment is relevant to the task at hand [10].

Eye-tracking techniques have been used in several studies to understand the interactions between cognitive processes and learning outcomes [11]. One of the interesting feature of the eye-tracking method is that it provides a way to track the encoding and attentional processes occurring during the learning phase [10].

In the literature, temporal eye-tracking measures have been found to be the most widely used in learning related studies [12]. Visit duration (i.e. cumulative duration of fixation within an area of interest) is considered as an indicator of the total amount of cognitive processing engaged with the fixated information [13]. Indeed, it has been suggested in the literature that learners give more attention and more time to complex problems than intermediate or easier problems [10, 14].

Expertise have been shown in the literature to have an impact on eye movements during learning [15–17]. It has been suggested that experts tend to attend more relevant features of a complex dynamic stimulus than novices, and that their attention remains focused on relevant areas [15]. This difference between experts and novices may also be found between individuals with smaller differences in expertise [18]. Besides, practice over a period of time has also been found to make individuals fixate faster and proportionally more on task relevant information [19].

Prior knowledge is also a factor that has been identified in the literature as having a significant impact on measures of eye movements. A study from Canham and Hegarty also suggest that newly acquired knowledge helped the learners focus their attention on task relevant knowledge and less on task irrelevant knowledge [20]. Their study consisted in two experiment in which the participants made inferences from weather maps before and after they received instruction about relevant meteorological principles [20]. Their results show that after receiving the instructions, the participants paid more attention to relevant information and less attention to irrelevant information [20]. We thus pose the following hypothesis:

H1: Vicarious training accelerates the attentional efficiency of the learner.

3 Method

3.1 Participants

To answer the research question, a between subject experimental design with two conditions has been chosen. The conditions assigned to each subject determined the order in which they received the enactive and vicarious trainings.

During the experiment, each subject had to follow both a 15 min vicarious and a 15 min enactive training session. In the first condition, the participants were first given the vicarious training. The vicarious training consisted of a 15 min demonstration video including a voiceover explaining the different principles that guided the creations of the different indicators and which of them were useful in the context of a given task. The participants then received the enactive part of the training in which they were asked to perform a task in the simulated business environment, i.e. ERPsim (Montréal, Canada) [21] using an online dashboard and a SAP GUI.

As illustrated in Fig. 1, the participants that were in the second condition followed the same protocol, but they received the enactive training first and then received the vicarious training. All the participants were asked to answer questionnaires assessing learning outcomes such as higher self-efficacy, objective knowledge, perceived difficulty of the training or satisfaction with the learning process [6, 7] before, between, and after the trainings.

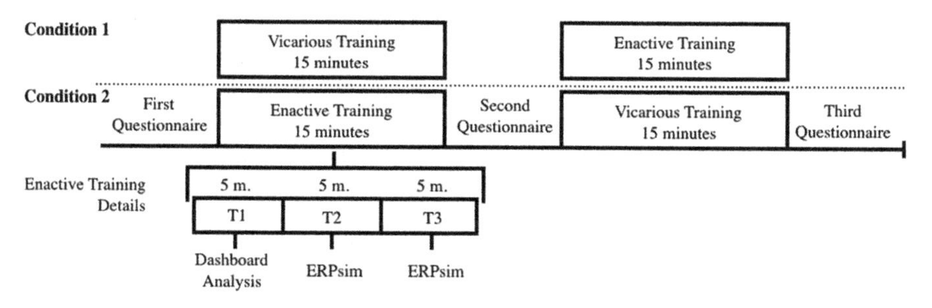

Fig. 1 Experiment timeline for both conditions

The experiment received the approval from Research Ethics Board of our institution. Thirty participants were recruited through a student panel that is comprised of university subjects between the ages of 18 and 35. Each experiment lasted 2 h and the participants were given a compensation of $30.

3.2 Design and Apparatus

The study was conducted using a platform to simulate a business environment called ERPsim [21]. The ERPsim games are known for providing a realistic business context in which trainees are using a real ERP system [i.e. SAP (Walldorf, Germany)] to manage their organization [4, 21–23]. Specifically, the "Logistics" variant of the game, has been used for this experiment to provide the enactive learning context. The task involving ERPsim required the participants to make business decisions regarding the quantity of each of their six products that were to be sent in the three available regions.

To help them execute the task in ERPsim, the participants were provided a self-refreshing dashboard that displayed data for sales, inventory, and financial performance as well as a memory aid for the different rules of the game. The visual portion dashboards were designed using the design principles from Stephen Few's books *Show me the Numbers* [24] and *Information Dashboard Design* [25]. The Dashboard contained four types of indicators: useful information, well presented; useful information, badly presented; useless information, well presented; useless information, badly presented. In total, four versions of the dashboard were produced to randomize the disposition of the indicators in the screen.

3.3 Instrumentation

3.3.1 Attentional Efficiency

We define the attentional efficiency as the learner's ability to identify and process rapidly relevant information from the dashboards. The participant's attentional efficiency was assessed using the visual attention of the participant during the experiment. Visual attention is measured using eye-tracking devices (Tobii X60) monitoring participant's gaze. Visual attention provide an objective measure of what participants were considering in the dashboard to make decision [26]. Average visit duration was assessed for each type of indicator on the dashboard because it provides clear information on the time it takes the participants to process the information.

3.3.2 Learning Outcomes

Learning outcomes, and perceived difficulty of the training, were measured using questionnaires that were developed with the learning outcomes constructs from the literature [7]: (a) participant's learning (objective knowledge) was measured by true or false questions with a certitude component, (b) participant's perceived understanding of the dashboard design principles and (c) the dashboard self-efficacy construct were measured using adapted elements from the self-efficacy items from Hollenbeck and Brief [27] to fit the experimental context of business dashboards, (d) the participants capacity to apply knowledge of dashboard usage was assessed with items oriented towards the main topics covered in Few [24, 25], (e) the level of satisfaction with the learning process was measured using the scale from Green and Taber [28], (f) the perceived enterprise system management knowledge construct was measured using the items from Cronan et al. [23].

4 Results and Discussion

To assess the attentional efficiency used during the training, we analyzed the participant's eye tracking data gathered during the enactive training. The vicarious training was excluded to avoid a bias linked to the explanations attracting attention to specific regions of the screen. The 15 min of the enactive training was divided in three 5 min segments to be able to differentiate the ability of the participants at different points during the training. For the analysis, the indicators were regrouped into the four categories mentioned above and we regressed the average visit duration for each of the categories. One of the main purposes of a dashboard being to allow the user to quickly identify things that deserve attention and might require action [25].

The results indicate that there are no significant differences between the two conditions for average visit duration on the indicators presenting useful information well-presented. However, the condition 2 (enactive first) spent significantly more time on average every time they visited the indicators that presented either useless information or a badly presented information or both of them (see Table 1). For example, the participants that did not receive the vicarious training, stayed 0.5 s (coefficient value) longer ($p = 0.028^{**}$) than their counterparts every time they visited an indicators showing useful information badly represented.

This is interesting because even though they had not received the vicarious training, the participants from condition 1 (vicarious first) were still able to interpret the data from the good indicators as rapidly as those who had received the vicarious training. Also, after only 15 min of enactive training, participants from condition 2 (enactive first) managed to catch up those in condition 1 (vicarious first) by reducing the difference to insignificant levels in the average visit duration for the indicators showing useless information well presented. This shows that even though

Table 1 Summary of attentional efficiency results: Regressions of the average visit duration (s) by condition for each type of indicators at the beginning (T1) and the end (T3) of the enactive training

	T1		T3	
Indicator type	Coefficient (s)	P-value	Coefficient (s)	P-value
Useful information, well presented	–	–	–	–
Useful information, badly presented	0.50	0.028**	0.48	0.003***
Useless information, well presented	0.23	0.078*	–	–
Useless information, badly presented	0.29	0.009***	0.24	0.046**

The coefficient values represent the increase average visit duration for condition 2 (enactive first) participants
*P < 0.1; **P < 0.05; ***P < 0.01; ****P < 0.001

they start without prior knowledge of business dashboard principles, subject from condition 2 (enactive first) progressed quickly in the interpretation of the different types of indicators. We thus accept our hypothesis that vicarious training accelerates attentional efficiency, but we also observe that the effect seems to wear out as enactive learners are able to catch up rapidly. It is also interesting that the version of the dashboard did not have any significant impact on the attentional efficiency of the participants.

While the main objective of the paper was not to compare the learning outcomes, we observe that out of the learning outcomes measured, the order in which the participants received the training only had a significant impact on the of dashboard self efficacy and perceived enterprise system management knowledge. Precisely, the participants that started with the enactive training had a lower increase in self efficacy (Coef. 0.55, p-value = 0.079*), but had a higher increase in their perceived enterprise system management knowledge (Coef. 0.54, p-value = 0.079*). This means that there were no significant differences between the two groups for the increase in objective knowledge at the end of the experiment. This suggest that both groups progressed the same no matter the sequence in which the trainings were provided. There were also no significant differences between the two groups for the perceived difficulty of the training and the satisfaction with the training process at the end of the experiment.

5 Conclusion

Eye-tracking results indicate (visit duration) that the sequence in which vicarious and enactive training occurs has a significant impact on the visual search patterns of students. However, the participants that did not receive the vicarious training before the enactive one managed to interpret data from the good indicators as quickly as those who had received the vicarious training and the difference reduces quickly after less than 15 min of training.

Learning outcome results suggest that the order in which vicarious and enactive training are provided has no impact on objective knowledge acquisition, satisfaction and perceived difficulty of the training process, and on the trainee's perception of their own knowledge of dashboard concepts. However, participants who received the vicarious training first, had a higher increase in their dashboard self efficacy, but had a lower increase in their understanding of enterprise system management knowledge than the participants that started with the enactive training. This implies that providing vicarious training first could lead to a higher academic performance in time compared to training methods where the enactive training is provided first because the correlation between self-efficacy and performance is stronger after a time lapse from the beginning of the learning experience [29].

Thus, based on our results, instructors should begin with the vicarious training when using a combination of enactive and vicarious training to optimize objective knowledge acquisition and the feelings of self efficacy. We prioritize the acquisition of self efficacy because it has been suggested in the literature that it has an impact on future performance whereas such evidence has not yet been demonstrated for the enterprise system management knowledge construct.

Finally, we conclude that eye-tracking can be a useful tool for the study of learning process of IS user as it can detect how learners process certain material [17]. It could allow researchers to compare pedagogical scenarios and propose efficient training methods based on empirical data.

References

1. Gupta, S., Anson, R.: Do I matter? The impact of individual differences on a technology-mediated end user training process. J. Organ. End User Comput. (JOEUC) **26**(2), 60–79 (2014)
2. Charland, P., Léger, P.M., Cronan, T.P., Robert, J.: Developing and assessing ERP competencies: basic and complex knowledge. J. Comput. Inf. Syst. **56**(1), 31–39 (2016)
3. Parenteau, J., Chandler, N., Salam, R.L., Laney, D., Duncan, A.D.: Predicts 2015: power shift in business intelligence and analytics will fuel disruption (ID: G00270932). Retrieved from Gartner database, 21 Nov 2014
4. Léger, P.-M., Cronan, T., Charland, P., Pellerin, R., Babin, G., Robert, J.: Authentic OM problem solving in ERP context. Int. J. Oper. Prod. Manag. **32**(12), 1375–1394 (2012)
5. Léger, P.-M., Davis, F.D., Cronan, T.P., Perret, J.: Neurophysiological correlates of cognitive absorption in an enactive training context. Comput. Hum. Behav. **34**, 273–283 (2014)
6. Gupta, S., Bostrom, R.P., Huber, M.: End-user training methods: what we know, need to know. ACM SIGMIS Database **41**(4), 9–39 (2010)
7. Gupta, S., Bostrom, R.: An investigation of the appropriation of technology-mediated training methods incorporating enactive and collaborative learning. Inf. Syst. Res. **24**(2), 454–469 (2013). doi:10.1287/isre.1120.0433
8. Yi, M.Y., Davis, F.D.: Developing and validating an observational learning model of computer software training and skill acquisition. Inf. Syst. Res. **14**(2), 146–170 (2003)
9. Just, M.A., Carpenter, P.A.: A theory of reading: from eye fixations to comprehension. Psychol. Rev. **87**(4), 329–354 (1980). doi:10.1037/0033-295X.87.4.329
10. Hyönä, J.: The use of eye movements in the study of multimedia learning. Learn. Instr. **20**(2), 172–176 (2010). doi:10.1016/j.learninstruc.2009.02.013

11. Anderson, O.R., Love, B.C., Tsai, M.: Neuroscience perspectives for science and mathematics learning in technology-enhanced learning environments: editorial. Int. J. Sci. Math. Educ. **12** (3), 467–474 (2014). doi:10.1007/s10763-014-9540-2
12. Lai, M.L., Tsai, M.J., Yang, F.Y., Hsu, C.Y., Liu, T.C., Lee, S.W.Y., et al. (2013). A review of using eye-tracking technology in exploring learning from 2000 to 2012. Educ. Res. Rev. **10**, 90–115
13. Ozcelik, E., Karakus, T., Kursun, E., Cagiltay, K.: An eye-tracking study of how color coding affects multimedia learning. Comput. Educ. **53**(2), 445–453 (2009). doi:10.1016/j.compedu.2009.03.002
14. Lin, J.J., Lin, S.S.J.: Cognitive load for configuration comprehension in computer-supported geometry problem solving: an eye movement perspective. Int. J. Sci. Math. Educ. **12**(3), 605–627 (2014). doi:10.1007/s10763-013-9479-8
15. Jarodzka, H., Scheiter, K., Gerjets, P., van Gog, T.: In the eyes of the beholder: how experts and novices interpret dynamic stimuli. Learn. Inst. **20**(2), 146–154 (2010). doi:10.1016/j.learninstruc.2009.02.019
16. Mayer, R.E.: Unique contributions of eye-tracking research to the study of learning with graphics. Learn. Instr. **20**(2), 167–171 (2010). doi:10.1016/j.learninstruc.2009.02.012
17. van Gog, T., Scheiter, K.: Eye tracking as a tool to study and enhance multimedia learning. Learn. Instr. **20**(2), 95–99 (2010). doi:10.1016/j.learninstruc.2009.02.009
18. van Gog, T., Paas, F., van Merriënboer, J.J.G., Witte, P.: Uncovering the problem-solving process: cued retrospective reporting versus concurrent and retrospective reporting. J. Exp. Psychol. Appl. **11**(4), 237–244 (2005). doi:10.1037/1076-898X.11.4.237
19. Haider, H., Frensch, P.A.: Eye movement during skill acquisition: more evidence for the information-reduction hypothesis. J Exp Psychol Learn Mem Cognit **25**(1), 172–190 (1999). doi:10.1037/0278-7393.25.1.172
20. Canham, M., Hegarty, M.: Effects of knowledge and display design on comprehension of complex graphics. Learn. Instr. **20**(2), 155–166 (2010). doi:10.1016/j.learninstruc.2009.02.014
21. Léger, P.-M.: Using a simulation game approach to teach enterprise resource planning concepts. J. Inf. Syst. Educ. **17**(4), 441–447 (2006)
22. Léger, P.-M., Charland, P., Feldstein, H.D., Robert, J., Babin, G., Lyle, D.: Business simulation training in information technology education: guidelines for new approaches in IT training. J. Inf. Technol. Educ. **10**, 28–51 (2011)
23. Cronan, T.P., Léger, P.-M., Robert, J., Babin, G., Charland, P.: Comparing objective measures and perceptions of cognitive learning in an ERP simulation game. Simul. Gaming **43**(4), 461–480 (2012). doi:10.1177/1046878111433783
24. Few, S.: Show Me the Numbers: Designing Tables and Graphs to Enlighten, 2nd edn. Analytics Press, Burlingam, CA (2012)
25. Few, S.: Information Dashboard Design: Displaying Data for At-a-Glance Monitoring, 2nd edn. Analytics Press, Burlingam, CA (2013)
26. Riedl, R., Léger, P.-M.: Tools in NeuroIS research: an overview. In: Riedl, R., Léger, P.-M. (eds.) Fundamentals of NeuroIS, pp. 47–72. Springer, Berlin (2016)
27. Hollenbeck, J.R., Brief, A.P.: The effects of individual differences and goal origin on goal setting and performance. Organ. Behav. Hum. Decis. Process. **40**(3), 392–414 (1987)
28. Green, S.G., Taber, T.D.: The effects of three social decision schemes on decision group process. Organ. Behav. Hum. Perform. **25**(1), 97–106 (1980)
29. Honicke, T., Broadbent, J.: The influence of academic self-efficacy on academic performance: a systematic review. Educ. Res. Rev. **17**, 63–84 (2016)

The Relationship Between Visual Website Complexity and a User's Mental Workload: A NeuroIS Perspective

Ricardo Buettner

Abstract I report results from an experiment on the relationship between visual website complexity and users' mental workload. Applying a pupillary based workload assessment as a NeuroIS methodology, I found indications that a balanced level of navigation complexity, i.e., the number of (sub)menus, in combination with a balanced level of information complexity, is the best choice from a user's mental workload perspective.

Keywords NeuroIS • Eye-tracking • Mental workload • Pupil diameter • IS complexity • Website complexity • Navigation complexity • Information complexity

1 Introduction

In this work I report on results from a completed experiment, for which prior research-in-progress results were presented last year on the Gmunden Retreat 2015 [1].

Website/webpage complexity affects a user's mental workload [2], and Huang [3] identified the amount of information and the number of links as important attributes of a website's complexity. The problem from a website design perspective is how to strike a balance between the dilemma of a complex menu structure (a lot of menu links and submenus) but non-complex pieces of information or a non-complex menu structure (with fewer links/submenus) with a high amount of information (more complex). To evaluate this problem researchers need a convenient way to assess a user's mental workload. Determining a user's mental workload is often mentioned as a fundamental problem in IS research (e.g. [4, 5]) from various theoretical perspectives (e.g., cognitive load, task technology fit, job demands-resources), particularly in NeuroIS, e.g. [6–11].

In recent years very interesting results have emerged from NeuroIS in which efforts have been made to determine a user's mental workload based on objective

R. Buettner
FOM University of Applied Sciences, MIS-Institute, Munich, Germany
e-mail: ricardo.buettner@fom.de

F.D. Davis et al. (eds.), *Information Systems and Neuroscience*, Lecture Notes in Information Systems and Organisation 16, DOI 10.1007/978-3-319-41402-7_14

psychophysiological measurements [9–11]. IS scholars have already used pupillary based mental workload assessment using realistic experimental setups, e.g., route planning [12, 13], E-mail classification [12], decision support systems [14], and social networks [7, 15].

To the best of my knowledge there is no study that investigates the relationship between visual website complexity and a user's mental workload using psychophysiological measures - with one exception: The work of Wang et al. [2] investigated website complexity from a cognitive load perspective via eye-tracking technology. Using fixation count and fixation duration they found increased fixation counts, fixation durations and task completion times when performing simple tasks. Interestingly they did not analyze pupillary measures in order to evaluate mental workload.

For this reason I study the usage of three website variants with systematic manipulations of navigation and information complexity using eye-tracking based pupillary diameter responses. This work contributes to IS complexity research, and, in addition, it addresses a very practical problem for website designers.

2 Methodology

2.1 Applying the NeuroIS Guidelines

In order to clearly contribute to NeuroIS research and show strong methodological rigor, I strictly followed the NeuroIS guidelines established by vom Brocke et al. [16]. In particular, to assess prior research in the field of measuring mental workload as an important IS construct, a comprehensive literature review was conducted (cf. [17]). To base the experimental design adequately on solid research in related fields of neuroscience [16] I reviewed the fundamental anatomic mechanism of the pupillary dilation controlled by the vegetative nervous system and the key role of the Edinger-Westphal nucleus that is inhibited by mental workload and directly leads to a pupillary dilation. The methodology uses eye-tracking-based pupillometry as a well-established approach in physiology and psychology "widening the 'window' of data collection" ([18], p. 93). With this method, bio-data (i.e. pupil diameter) can be used to better understand mental workload as an IS construct (cf. guideline 4 of [16]). In comparison to other neuroscience tools eye-tracking-based pupillometry is the contact-free and efficient method of choice [19]. I applied the guidelines and standards from Duchowski [20] and the Eyegaze Edge™ manual.

2.2 Measurements

To capture the pupillary diameter, eye-tracking was performed using the binocular double Eyegaze Edge™ System eye-tracker paired with a 19″ LCD monitor (86 dpi) set at a resolution of 1280 × 1024, whereby the eye-tracker samples the pupillary diameter at a rate of 60 Hz for each eye separately.

2.3 Stimuli

Following [21] I manipulated website visual complexity via the number of links in the menu structure (resp. submenus). According to Wang et al. [2] I choose three contrary but balanced levels for navigation and information complexity. Navigation complexity was manipulated by the (sub)menu structure (low: 3 menus; average: 3 × 3 (sub)menus; high: 3 × 3 × 3 (sub)menus). Information complexity was manipulated by content/text partitioning. All three variants (system A, B, C; see Fig. 1) contained the same content/information in summary, but I divided this content into (sub)menu-specific pieces of information. Luminescence levels of the three systems variants were checked (perceived relative luminescence $L_A = 0.5156$, $L_B = 0.5157$, $L_C = 0.5209$).

Please note that I directly tested objective website complexity, since perceived website complexity correlated only medially with objective website complexity ($r = 0.3$ according to [22], p. 515; Fig. 2).

2.4 Description of the Test Procedure and Data Cleansing

The participants in the experiment had to perform nine distinctive search tasks - three for each system. In order to counter-balance the design, the test order of the systems

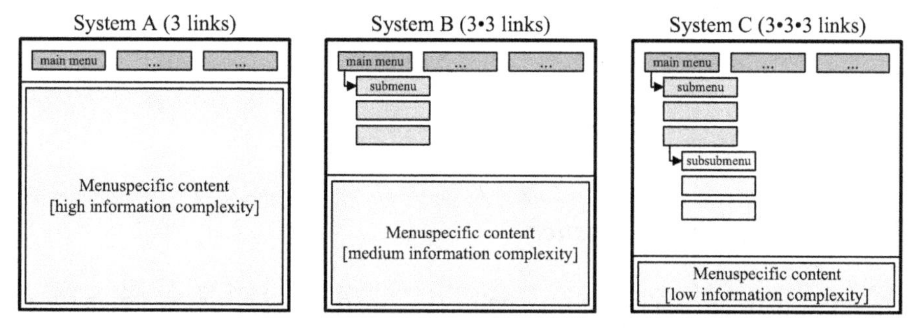

Fig. 1 Conceptualized website complexity (System A: low menu complexity - high information complexity; System B: average menu complexity - average information complexity; System C: high menu complexity - low information complexity)

Fig. 2 Exemplary screenshot of system A

(A, B, C) was randomized. In addition, for every test system (A, B, C) three of the nine search tasks were randomly assigned.

Prior to all data collection, each test participant is welcomed by the experimenter (supervisor of the experiment). After that the participant has to fill out a consent form and also a questionnaire with demographics (stage 1). In stage 2, I take the necessary precautions for the experiment, for which I make use of the eye-tracker. Hence, the eye-tracker is calibrated. In stage 3, the experiment starts with the first of three search tasks the participant has to accomplish. At the end of every tested system (A, B, C), perceived task demands and effort were captured via a NASA TLX questionnaire.

Only naturally determined artifacts, e.g., by eye-blinks, were deleted.

3 Results

3.1 Sample Characteristics

The 45 participants were aged from 22 to 45 years (M = 28.6, S.D. = 4.7). 24 persons were female, 21 male.

Table 1 Mean of pupillary diameters in relation to system variant

System	PD [mm]	
	Left eye	Right eye
System A	3.402	3.451
System B	3.384	3.435
System C	3.428	3.479

Table 2 2-sided significance test (p-value, t-statistics), effect sizes of different system variants

Comparison	Significance test results	
	Left eye	Right eye
System A vs. system B	$p < 0.1$, $t = 1.96$, $d = 0.29$ (small)	$p < 0.1$, $t = 1.79$, $d = 0.27$ (small)
System A vs. system C	$p < 0.05$, $t = 2.49$, $d = 0.37$ (small)	$p < 0.05$, $t = 2.63$, $d = 0.39$ (small)
System B vs. system C	$p < 10^{-5}$, $t = 5.05$, $d = 0.75$ (medium)	$p < 10^{-4}$, $t = 4.79$, $d = 0.71$ (medium)

3.2 Relationship Between Complexity and a User's Mental Workload

I found clear pupil diameter differences between the three system variants (Table 1) which were all significant at small to medium effect sizes (Table 2).

4 Discussion, Limitations and Future Research

From a mental workload perspective, the system B is the model of choice since the pupillary based mental workload indicator is lowest for this system variant (small to medium Cohen's d effect sizes). That means for the practical website design perspective that complex menu structures with a lot of menu links and submenus or a design without any submenus should be avoided. Instead, designers should use a balanced combination of submenus (navigation complexity) and text (more information complexity).

From a theoretical point of view this work contributes to IS complexity research. The results indicate that pure navigation complexity (i.e., the number of (sub) menus) or pure information complexity (text) is problematic from a mental workload perspective.

This work has some limitations. The right pupil diameters were slightly larger than for left eyes due to small differences in illumination from the ceiling lights. However, these lighting conditions were kept constant for all participants. In addition, despite designing the experiment as a search task, it cannot be excluded that some participants merely browsed the website. Because there is a mental workload difference between browsing and searching [23], the generalizations of the findings are limited. Furthermore, despite controlling the brightness between the three system variants through the most balanced use of darker text against brighter text background, the variant C is one percent brighter than A and B. However this higher brightness naturally leads to smaller pupils, but not to the larger pupils I found in the experiment. This further strengthens the mental workload effect in system variant C.

In an extended version of this paper I will report on triangulated NASA TLX evaluations, differences in mouse clicks and mouse scrolls as well as results from electrodermal activity assessments. Future work should combine pupillary-based mental workload assessment with novel heat-mapping techniques [24, 25].

Acknowledgements I would like to thank Christiane Lange for laboratory assistance and the reviewers, who provided very helpful comments on the refinement of the paper. This research is funded by the German Federal Ministry of Education and Research (03FH055PX2).

References

1. Buettner, R.: Investigation of the relationship between visual website complexity and users' mental workload: A NeuroIS perspective. In: Information Systems and Neuro Science, vol. 10 of LNISO, pp. 123–128. Gmunden, Austria (2015)
2. Wang, Q., Yang, S., Liu, M., Cao, Z., Ma, Q.: An eye-tracking study of website complexity from cognitive load perspective. Decis Support Syst **62**, 1–10 (2014)
3. Huang, M.-H.: Designing website attributes to induce experiential encounters. Comput. Hum. Behav. **19**(4), 425–442 (2003)
4. Stassen, H.G., Johannsen, G., Moray, N.: Internal representation, internal model, human performance model and mental workload. Automatica **26**(4), 811–820 (1990)
5. Johannsen, G., Levis, A.H., Stassen, H.G.: Theoretical problems in man-machine systems and their experimental validation. Automatica **30**(2), 217–231 (1992)
6. Buettner, R.: Analyzing mental workload states on the basis of the pupillary hippus. In: NeuroIS '14 Proc., p. 52 (2014)
7. Buettner, R., Sauer, S., Maier, C., Eckhardt, A.: Towards ex ante prediction of user performance: A novel NeuroIS methodology based on real-time measurement of mental effort. In: HICSS-48 Proc., pp. 533–542 (2015)
8. de Guinea, A.O., Titah, R., Léger, P.-M.: Explicit and implicit antecedents of users' behavioral beliefs in information systems: A neuropsychological investigation. J Manag Inform Syst **30** (4), 179–210 (2014)
9. Dimoka, A., Pavlou, P.A., Davis, F.D.: NeuroIS: The potential of cognitive neuroscience for information systems research. Inform Syst Res **22**(4), 687–702 (2011)
10. Dimoka, A., Banker, R.D., Benbasat, I., Davis, F.D., Dennis, A.R., Gefen, D., Gupta, A., Ischebeck, A., Kenning, P.H., Pavlou, P.A., Müller-Putz, G., Riedl, R., vom Brocke, J., Weber, B.: On the use of neurophysiological tools in IS research: Developing a research agenda for NeuroIS. MIS Q **36**(3), 679–A19 (2012)
11. Riedl, R., Banker, R.D., Benbasat, I., Davis, F.D., Dennis, A.R., Dimoka, A., Gefen, D., Gupta, A., Ischebeck, A., Kenning, P., Müller-Putz, G., Pavlou, P.A., Straub, D.W., vom Brocke, J., Weber, B.: On the foundations of NeuroIS: Reflections on the Gmunden retreat 2009. Comm Assoc Inform Syst **27**, 243–264 (2010)

12. Iqbal, S.T., Adamczyk, P.D., Zheng, X.S., Bailey, B.P.: Towards an index of opportunity: Understanding changes in mental workload during task execution. In: CHI '05 Proc., pp. 311–320 (2005)
13. Bailey, B.P., Iqbal, S.T.: Understanding changes in mental workload during execution of goal-directed tasks and its application for interruption management. ACM Trans Comput Hum Interact **14**(4), 1–28 (2008). Article 21
14. Buettner, R.: Cognitive workload of humans using artificial intelligence systems: Towards objective measurement applying eye-tracking technology. In: KI 2013 Proc., ser. LNAI, vol. 8077, pp. 37–48 (2013)
15. Buettner, R., Daxenberger, B., Eckhardt, A., Maier, C.: Cognitive workload induced by information systems: Introducing an objective way of measuring based on pupillary diameter responses. In: Pre-ICIS HCI/MIS 2013 Proc., paper 20 (2013)
16. vom Brocke, J., Liang, T.-P.: Guidelines for neuroscience studies in information systems research. J Manag Inform Syst **30**(4), 211–234 (2014)
17. vom Brocke, J., Simons, A., Niehaves, B., Riemer, K., Plattfaut, R., Cleven, A.: Reconstructing the giant: On the importance of rigour in documenting the literature search process. In: ECIS '09 Proc., pp. 2206–2217 (2009)
18. Goldinger, S.D., Papesh, M.H.: Pupil dilation reflects the creation and retrieval of memories. Curr Dir Psychol Sci **21**(2), 90–95 (2012)
19. Laeng, B., Sirois, S., Gredebäck, G.: Pupillometry: A window to the preconscious? Perspect Psychol Sci **7**(1), 18–27 (2012)
20. Duchowski, A.T.: Eye tracking methodology: Theory and practice, 2nd edn. Springer, London (2007)
21. Deng, L., Poole, M.S.: Affect in web interfaces: A study of the impacts of web page visual complexity and order. MIS Q **34**(4), 711–A10 (2010)
22. Nadkarni, S., Gupta, R.: A task-based model of perceived website complexity. MIS Q **31**(3), 501–524 (2007)
23. Hong, W., Thong, J.Y., Tam, K.Y.: The effects of information format and shopping task on consumers' online shopping behavior: A cognitive fit perspective. J Manag Inform Syst **21**(3), 149–184 (2004)
24. Eckhardt, A., Maier, C., Hsieh, J.J., Chuk, T., Chan, A.B., Hsiao, J.H., Buettner, R.: Objective measures of IS usage behavior under conditions of experience and pressure using eye fixation data. In: ICIS 2013 Proc. (2013)
25. Georges, V., Courtemanche, F., Sénécal, S., Baccino, T., Léger, P.-M., Frédette, M.: Measuring visual complexity using neurophysiological data. In: Information Systems and Neuro Science, vol. 10 of LNISO, pp. 207–212. Gmunden, Austria (2015)

Studying the Creation of Design Artifacts

Christopher J. Davis, Alan R. Hevner, and Barbara Weber

Abstract As software and information systems (IS) increase in functional sophistication, perceptions of IS quality are changing. Moving beyond issues of performance efficiency, essential qualities such as fitness for purpose, sustainability, and overall effectiveness become more complex. Creating software and information systems represents a highly interconnected locus in which both the generative processes of building design artifacts and articulating constructs used to evaluate their quality take place. We address this interconnectedness with an extended process-oriented research design enabling multi-modal neurophysiological data analyses. We posit that our research will provide more comprehensive assessments of the efficacy of design processes and the evaluation of the qualities of the resulting design artifacts.

Keywords IS design • Creating design artifacts • Personal Construct Psychology (PCT) • Event-related potential (ERP) • Eye-tracking • Eye fixation related potential (EFRP) • Electroencephalography (EEG) • Interaction Logging • Repertory Grid Analysis (RGA)

1 Introduction

Creative design activities are central to all applied engineering disciplines. The information systems (IS) field since its advent has the principal objective of designing, building, and evaluating systems to solve complex business problems.

C.J. Davis (✉)
University of South Florida, Saint Petersburg, FL, USA
e-mail: davisc@mail.usf.edu

A.R. Hevner
University of South Florida, Tampa, FL, USA
e-mail: ahevner@usf.edu

B. Weber
Technical University of Denmark, Copenhagen, Denmark

University of Innsbruck, Innsbruck, Austria
e-mail: bweb@dtu.dk

F.D. Davis et al. (eds.), *Information Systems and Neuroscience*, Lecture Notes in Information Systems and Organisation 16, DOI 10.1007/978-3-319-41402-7_15

IS as composed of inherently mutable and adaptable hardware, software, telecommunications, and human interfaces provide many unique and challenging design problems that call for new, creative ideas and discovery. IS artifacts are implemented within an application context for the purpose of improving the effectiveness and efficiency of that context. The utility of the artifact and the characteristics of the application—its work systems, its people, and its development and implementation methodologies—together determine the extent to which that purpose is achieved. Researchers produce new ideas that enhance generative capacity [1] and improve the ability of human organizations to adapt and succeed in the presence of changing environments. These generative ideas are then communicated as knowledge to the various IS communities [2].

The IS design environment is characterized by significant (and increasingly complex) generative opportunities presented by the diversification of rapidly evolving technologies in terms of development focus and medium (e.g. 'wrappers' for legacy systems; XaaS; cloud/virtualization; mobile apps) and agility (e.g. speed of creation and deployment). Creative design activities are supported by a plethora of representational methods and tools. With all of this richness of creative design opportunities and enabling creative infrastructures, IS researchers still struggle to understand the process of creating artifacts in IS design—and how to measure their quality. Thus, we pose the research question: Has the diversification of functionalities, development environments, tools (e.g. representational languages) etc. expanded the range of criteria that are used to guide the creative design process and improved our ability to judge the quality of design artifacts? In other words, have new understandings and theories of creating design artifacts emerged in response to advances in design processes and tool evolution and their innovative use? We posit that the answer is *Yes*. The remainder of this paper presents a comprehensive research design and protocols that enable emergent characteristics of design quality to be identified and articulated, providing insight into the process of creating design artifacts. We triangulate a rich array of previously inchoate empirical data sources to rigorously address the research question set out above. This paper extends research proposed by Davis and Hevner [3] on how designers employ visual syntax in the process of creating IS design artifacts.

2 The Creation of Design Artifacts

Design is both a process and a product. It describes the world as acted upon (processes) and the world as sensed (artifacts). In this research the phenomena of interest are the creation process and the quality assessment of the created design artifacts. As we have noted previously [4] this view of design supports a problem-solving paradigm that continuously shifts perspective between design processes and designed artifacts for the same complex problem. The design process is a sequence of activities tapping a range of expertise that produces an innovative product (i.e., the design artifact). The perceived qualities of the artifact enable evaluation which provides feedback, leading to a better understanding of the design challenge and, in turn, to improvement of the qualities of both the product and the design process.

Hevner, Davis, Collins, and Gill [4] propose a 2 × 2 model of the design process from the perspective of neuroscience. The x-axis distinguishes the External (Task) Environment from the Internal (Cognitive) Environment; the y-axis separates the Problem Space from the Solution Space. In this research, we focus on the iterations of observation and generation of candidate designs that advance the design process from the upper-right internal problem space to the lower-right internal solution space, creating new, effective candidate solutions in response to the requirements in the problem space—the process of creation. This flow enables creative design events—decisions and actions—to be captured and analyzed in real-time. The form and immediacy of the post-task analyses articulate the constructs that guide design events and evaluation of the quality of the candidate solution artifact. Our research framework overcomes the limitations of instruments such as the Remote Associates Test and the Alternative Uses Task: their emphasis on comprehension limits their capacity to accommodate emerging, i.e. not yet fully known, cognitive constructs.

The specific questions addressed here concern the characteristics used to guide the creation of conceptual models (e.g. UML diagrams). We argue that this generative component of design differs substantially from the primarily comprehension based tasks that have guided prior research (e.g. [5–8]). Our goal is to identify the constructs used by designers to articulate the qualities that guide creating and evaluating designs. Our prior work [9] shows that the combination of neurophysiological data—particularly those elicited using the EFRP and ERP protocols—and interaction logs provide an authoritative basis to identify design decisions and the quality characteristics driving them. In this paper, we present a comprehensive research design and protocol, in Research Design and Research Protocol sections, respectively, that enable neurophysiological data and interaction logs to be synchronized. This composite data set provides the basis for eliciting the constructs undergirding design choices and design actions, i.e., interactions with the design platform.

3 Research Design

The build and evaluate cycles of design are typically iterated a number of times before the final (use) artifact is released into an application context for further testing and assessment through field study. During the process of creating and refining design artifacts the researcher must be cognizant that both the design process and the design artifact evolve as the research progresses. Kelly's [10] Personal Construct Theory (PCT) accommodates the socio-cognitive milieu that characterizes the co-evolution of the process of creating design artifacts and the constructs that guide their evaluation.

PCT [10] was developed as a framework for understanding how people made sense of the world around them. Kelly argued that people act as 'lay epistemologists' [11] in their attempts to order and interpret their experiences, categorizing and discriminating between them. In this way, individuals develop systems of interrelated personal constructs that enable them to anticipate the consequences of their own actions—and interpret the actions of others. PCT allows researchers to tap into the mental models used by individual designers to frame and articulate the world as acted upon and the world as sensed. PCT is particularly well suited to our phenomenon of interest—the qualities guiding the process of creating and evaluating design artifacts.

PCT is translated into practice using a cognitive mapping technique called Repertory Grid Analysis (RGA). Individuals' perceptions of similarity and difference are elicited, tapping into their theories of how the world operates. Use in its 'minimum context' form exploits the potential of RGA to directly tap designers' perceptions. This maintains the integrity of "the complex interconnectedness of the self and its social surroundings" [12, p. 345] providing two substantial benefits. First, it accommodates a wide range of empirical data types during elicitation of perceptions of the design process and product, including intermediate versions of the design artifact in the processes of building artifacts and evaluating progress. This enables design decisions that drive evolution of the design artifact from one version to another to be articulated directly by their creator. Second, relying on PCT to guide analysis of these constructs provides more ready and authentic construing of construction processes by members of the design community that we study. This responds directly to calls from the NeuroIS community for robustness and rigor in the application of neuro techniques [13]. The need for a more a comprehensive research protocol in order to accommodate the longer-term (multi-episodic) series of cognitive 'events' that generative tasks associated with creating software design artifacts entail is highlighted by Weber et al. [9]. The protocol set out below strives to maintain the 'interconnectedness' of designer and design [12] in a workplace setting, significantly increasing the immersion of the research into the design context—and authenticity of the analyses.

4 Research Protocol

Figure 1 outlines the research protocol and shows a three-phase 'immersion' into the complex interconnectedness of design environments. In order to exploit the capacities of PCT and RGA, identification of cognitive 'events' indicating design decisions and reconstruction of the design artifact before and after the design decision are required. In Phase 1 the designer works on a design task, i.e., creates a design artifact using a design platform. During this phase we rely on multi-modal data collection: we conduct neuro-physiological measurements, collect EEG data and instrument the design platform to simultaneously gather data on the design

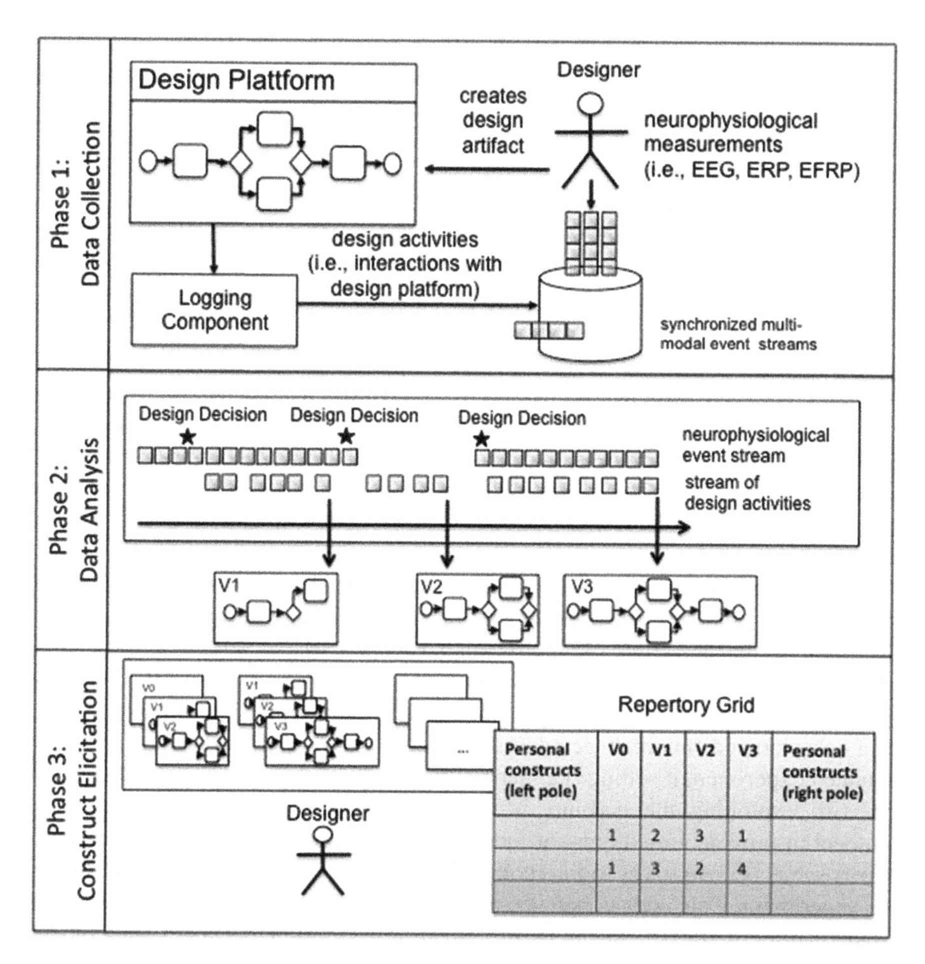

Fig. 1 Immersive analysis protocol

activities, i.e., interactions with the design platform. Data are synchronized using timestamps.

In Phase 2 we address the transition of designers from creating to evaluating: the events that interconnect designer and design. The stream of neurophysiological data is used to identify relevant cognitive events, i.e., the points in time when decisions took place, using the ERP and EFRP protocols with the EEG data. Timestamps are used to extract an intermediate model associated with each cognitive event from the stream of design activities [14–17] providing a secure basis for data triangulation.

In Phase 3 a series of steps is used to elicit personal constructs. Firstly, intermediate model versions are presented to the designer three at a time: their similarities and differences are used to articulate the 'poles' of a dimension that differentiate them. The poles provide the labels for a row in the repertory grid.

This step is repeated until all version combinations have been exhausted and/or repetition occurs. The columns of the grid are labelled using the (numbered) model versions—the design task 'elements' that the grid represents. The rows are also numbered, 1 indicating the left-side label and 5 indicating the right-side label, providing a simple Likert scale for later use.

The (subject) designer completes the next step alone: working one row at a time, the designer rates each model version (element) using the 1–5 scale. The completed grid is analyzed using two-dimensional (spaced-focused) cluster analysis to re-order the grid and add dendograms (crow's feet) to illustrate the linkages or clustering of the constructs.

The final step uses a protocol called 'talkback' to guide reflection on the spaced-focused grid. The designer articulates the qualities that characterize the most closely linked constructs: these are recorded by the researcher. This interview phase is iterative, moving from the smallest clusters 'outwards' to articulate more universal constructs. The system of inter-related constructs that represent the subject designer's view of the essential characteristics of design is captured in this phase.

The protocol balances the subject-centricity offered by PCT—essential to elicit the constructs guiding the process of creating and evaluating design artifacts—with the rigor of the detection of intermediate versions of the design artifact from interaction logs, ERP, EFRP and RGA protocols. There are three specific benefits: (i) accommodation of a range of complementary data types; (ii) maintenance of data cohesion and 'interconnectedness' by eliciting and analyzing them in a well-bounded experimental setting (i.e. generative design in the world of the designer); and (iii) exploiting the capacity of minimum context RGA and the talkback protocol to provide a medium—or 'conversational technology' [18]—that enables expert subjects *themselves* to interpolate the disparate data types generated during the experiment. This adds coherence and accommodates the longevity and cumulative nature of design, enabling investigation of design decisions and model interactions in near to real-time. Further, the RGA protocols mitigate limitations of techniques such as talk-aloud, removing the risks associated with loss of recall and interference with neurophysiological data gathering. The re-connection of designer and design process through RGA overcomes these limitations, enabling disparate data types to be orchestrated in real-time to provide new insights into the design process as it unfolds in the work setting.

We anticipate that prompt use of RGA to re-immerse the subject in their design experience will enable the constructs used to create and evaluate design artifacts to be readily and reliably elicited. Prior use of these protocols has been revelatory to both researcher and subjects, identifying constructs that had previously neither been anticipated or articulated [11]. Such emergent constructs will, we envisage, enable the range of criteria used to assess the utility and other qualities of design processes and artifacts to be significantly expanded.

5 Discussion and Future Directions

The research design is novel: the phenomena of interest—the design processes and the qualities of the artifacts they generate—have proved elusive. This novelty is complemented by focus on the creative transitions between building and evaluating, offering new insight into the qualities that drive the processes of creating and selecting candidate design artifacts.

The combination of process-orientation with multi-modal data gathering has significant benefits and implications for both theory and practice. The immersive analysis substantially extends and enriches assessment of design processes and artifacts. Fuller understanding of the qualities that guide design decisions during the creation of design artifacts will also provide useful advice for practicing designers and trainers.

PCT maintains the interconnectedness of designer and design and moves towards a more tractable definition of creation—the 'process of creating design artifacts', overcoming the limitations of prior neuro and NeuroIS studies [13, 14]. For instance, research into conceptual modeling is dominated by comprehension-oriented tasks (e.g., [19–22]) rather than conceptual modeling tasks (e.g., [23]) and research settings applying neuro-physiological tools favor static diagrams over (dynamic) processes (e.g., [5–8]). Moreover, in a more general context, most of the studies applying neuro-physiological tools in IS are stimulus-response tasks rather than generative—addressing short-term experiments (episodes) rather than longer-term processes [14, p. 107].

The paper also serves as an invitation to the NeuroIS community to provide feedback on the feasibility and viability of the research design and to explore collaborative means of achieving the critical mass of resources required to bring it to fruition.

References

1. Avital, M., Te'eni, D.: From generative fit to generative capacity: Exploring an emerging dimension of information systems design and task performance. Inform Syst J **19**(4), 345–367 (2009)
2. Gregor, S., Hevner, A.: Positioning and presenting design science research for maximum impact. MIS Q **37**(2), 337–355 (2013)
3. Davis, C., Hevner, A.: Neurophysiological analysis of visual syntax in design. In: Information Systems and Neuroscience: Gmunden Retreat on NeuroIS, pp. 99–106. Springer, Gmunden, Austria (2015)
4. Hevner, A., Davis, C., Collins, R., Gill, T.: A neurodesign model for IS research. Inf. Sci. Int. J. Emerg. Transdiscipl. **17**, 103–132 (2014)
5. Mendling, J., Strembeck, M., Recker, J.: Factors of process model comprehension—Findings from a series of experiments. Decis Support Syst **53**(1), 195–206 (2012)
6. Figl, K., Recker, J., Mendling, J.: A study on the effects of routing symbol design on process model comprehension. Decis Support Syst **54**(2), 1104–1118 (2013)

7. Figl, K., Mendling, J., Strembeck, M.: The influence of notational deficiencies on process model comprehension. J Assoc Inform Syst **14**(6), 312–338 (2013). Article 1
8. Recker, J., Reijers, H.A., Van de Wouw, S.G.: Process model comprehension: the effects of cognitive abilities, learning style, and strategy. Comm Assoc Inform Syst **34**(9), 199–222 (2014)
9. Weber, B., Pinggera, J., Neurater, M., Zugal, S., Martini, M., Furtner, M., Sachse, P. and Schnitzer, D.: Fixation patterns during process model creation: Initial steps toward neuro-adaptive process modeling environments. In: Proceedings of the 49th Hawai'i International Conference on Systems Sciences (HICSS49) Kauai, Hawai'i, pp. 600–609 (2016)
10. Kelly, G.: The Psychology of Personal Constructs. Norton, London (1955)
11. Davis, C., Hufnagel, E.: Through the eyes of experts: A socio-cognitive perspective on the automation of fingerprint work. MIS Q **31**(4), 681–703 (2007)
12. Neimeyer, G., Neimeyer, R.: Relational trajectories: A personal construct contribution. J Soc Pers Relat **2**, 325–349 (1985)
13. Riedl, R., Davis, F., Hevner, A.: Towards a NeuroIS research methodology: Intensifying the discussion on methods, tools and measurement. J Assoc Inform Syst **15**, i–xxxv (2014)
14. Riedl, R., Léger, J.-P.: Fundamentals of NeuroIS: Information Systems and the Brain. Studies in Neuroscience, Psychology, and Behavioral Economics. Springer, Berlin (2016)
15. Müller-Putz, G., Riedl, R., Wriessnegger, S.: Electroencephalography (EEG) as a research tool in the information systems discipline: Foundations, measurement and applications. Comm Assoc Inform Syst **37**(46), 911–948 (2015)
16. Pinggera, J., Zugal, S., Weidlich, M., Fahland, D., Weber, B., Mendling, J., Reijers, H.: Tracing the process of process modeling with modeling phase diagrams. In: Proc. ER-BPM '11, pp. 370–382 (2012)
17. Pinggera, J., Zugal, S., Weber, B.: Investigating the process of process modeling with cheetah experimental platform. In: Proc. ER-POIS '10, pp. 13–18 (2010)
18. Thomas, L., Harri-Augstein, S.: Self-Organized Learning: Foundations of a Conversational Science for Psychology. Routledge & Kegan-Paul, London (1985)
19. Matsuo, N., Ohkita, Y., Tomita, Y., Honda, S., Matsunaga, K.: Estimation of an unexpected-overlooking error by means of the single eye fixation related potential analysis with wavelet transform filter. Int J Psychophysiol **40**(3), 195–200 (2001)
20. Takeda, Y., Sugai, M., Yagi, A.: Eye fixation potentials in a proof-reading task. Int J Psychophysiol **40**(3), 181–186 (2001)
21. vom Brocke, J., Riedl, R., Léger, P.: Application strategies for neuroscience in information systems design science research. J Comput Inform Syst **53**(3), 1–13 (2013)
22. Recker, J., Safrudin, N., Rosemann, M.: How novices design business processes. Inf. Syst. **37** (6), 557–573 (2012)
23. Hungerford, B., Hevner, A., Collins, R.: Reviewing software diagrams: A cognitive study. IEEE Trans Software Eng **30**(2), 82–96 (2004)

Bridging Aesthetics and Positivism in IS Visual Systems Design with Neuroscience: A Pluralistic Research Framework and Typology

Daniel Peak, Victor Prybutok, Bin Mai, and Thomas Parsons

Abstract In this paper, we proposed a neuroscience-based general framework for IS visual systems design that uses neuroscience as the bridge that systematically integrates the principles of design aesthetics and principles of positivistic functionalities into a comprehensive model. Based on our general conceptual model, we also provide a detailed typological framework for the practical IS visual systems design.

Keywords Neuroscience • NeuroIS • Aesthetics • Positivism • Systems design • Methodology • Topological framework

1 Introduction

Information systems (IS) visual system design is an important component of the general concept of information design. As [1] point out, "in order to satisfy the information needs of the intended receivers, information design comprises analysis, planning, presentation and understanding of a message—its content, language and form... a well-designed information set, with its message, will satisfy aesthetic, economic, ergonomic, as well as subject matter requirements" (p. 19). In the past decades, visual system design has become an increasingly relevant topic in IS, as evidenced by a plethora of relevant studies [2–6]. Generally, in IS there are two research streams that address visual system design [7]: one holds that interface usability and functionality is the key, emphasizing a behavioral or cognitive focus [8–12]; the other research stream contends that attention to hedonic aspects of human-computer interaction, with human needs such as emotion, affect, and experience is important [13–15].

D. Peak • V. Prybutok • B. Mai (✉) • T. Parsons
University of North Texas, Denton, TX, USA
e-mail: daniel.peak@unt.edu; victor.prybutok@unt.edu; bin.mai@unt.edu; thomas.parson@unt.edu

F.D. Davis et al. (eds.), *Information Systems and Neuroscience*, Lecture Notes in Information Systems and Organisation 16, DOI 10.1007/978-3-319-41402-7_16

Traditionally in the IS field, the visual system design is studied through the positivist paradigm with a functional approach. According to this approach, the IS system design methodology focuses on the user's cognitive perceptions of the IS visual system design artifacts, and the factors influencing such perceptions. The research papers with this approach usually utilize the terms "design," "development," "dimension," "construct," "characteristic," "variable," and "factor". There are three positivist dimensions that form the premises of IT systems development [16]: factors of systems development, user outcomes, and owner value outcomes [17, 18].

Historically, there exists another separated paradigm for visual system design: the aesthetics of information design that focuses on the visual impression that "is both instantaneous and persistent in memory" [19]. The vocabulary used by researchers with this approach usually includes terms such as "visual," "aesthetic," "sensory," "interactive," and "beauty". The three aesthetic dimensions that form the fundamental premises of the visual design discipline and well-established in the literature are: elements of visual design, principles of visual design, and factors of visual composition [20–24].

Researchers have argued for a systematic approach that can integrate these two distinct, and currently mostly-independent paradigms, and call for a comprehensive framework that illustrates the IS visual systems design with both functionality and aesthetic appeals [25, 26]. In this paper, we propose an innovative framework to integrate these two disparate paradigms: we use neuroscience as an underlying connection bridging aesthetics design principles with positivistic design principles. On one hand, we illustrate the design model for Neuro-aesthetics, and on the other hand, we also describe the NeuroIS model for system design. Then we show that these two approaches can be seamlessly integrated into one comprehensive framework for IS visual systems design. In addition, based on our proposed general framework, we develop a typological model for the practical application of visual systems design.

2 Neural Foundation for IS Design Science

In IS field, design science research refers to the stream of research that focuses on the "development of novel IT artifacts, organizational development, and theory building" [27, 28]. Traditionally this area has provided structure and direction to the design of IS components, and identified factors that influence the processes [29].

One field that holds great promise in contributing to the above area of IS design science is neuroscience, which is the science that explains how human brain works. NeuroIS is the emerging field that focuses on the integration of neuroscience and IS [30, 31]. And especially in IS design science, for which human cognitive perceptions and decision making play a critical role, neuroscience provides great potential in contributing to the enhancement of IS design science methodology. The intersection between NeuroIS and IS design science has been investigated extensively since 2010 [32–35].

In this paper, we adopt [29]'s NeuroIS Design Science Model to illustrate the neural foundation for IS design science. Their model focuses primarily on three steps of interaction between human designers' neural activities and IS artifacts, and can briefly described as following:

- In the first step, the designers measure the artifact efficiency using neuroscience and generate neurophysiologic data. The step can be closely mapped to the dimension of "factors of systems development" in the positivist premise of IT systems development.
- In the second step, new or improved IS artifacts and interfaces are designed, developed and built using the measurement data generated from the first step, in order to enhance the efficiency of the system. The system efficiency can be measured through neurophysiologic values such as cognitive load, stress, and working memory, etc. This step can be closely mapped to the dimension of "user outcomes" in the positivist premise of IT systems development.
- Finally, the outcomes of the new or improved IS artifacts are tested and evaluated using neural cognitive methods to provide an overall evaluation to the system owners. This step can be closely mapped to the dimension of "owner value outcomes" in the positivist premise of IT systems development.

Therefore, the neural foundation for IS design science can thus be modeled by Fig. 1.

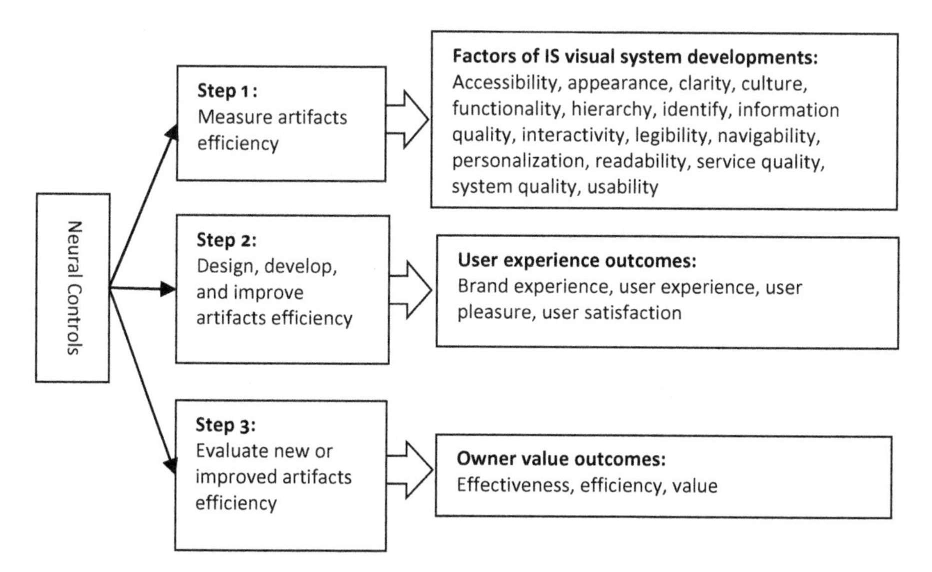

Fig. 1 The model of neural foundation for IS system design science

3 Neural Foundation for Visual Aesthetic Perception

As [36] pointed out, there are parallels between properties of art, more specifically, the principles of perception, and organizational principles of the brain. Production and perception of art ought to conform to principles of neural organizations. Neuroaesthetics is an emerging area, and various researchers have proposed models of visual aesthetic appreciation, perception, and judgments based on cognitive neuroscience. Two main neuroaesthetic models are [37] and [38].

In [37]'s model, the neural processes for aesthetic interaction starts with an affective state invoked by a design artifact, and goes through steps of perceptual analysis, implicit memory integration, explicit classification, cognitive mastering, and evaluation, before it reach the stage of aesthetic judgment and aesthetic emotion. During each of the steps, the outcomes of those steps are significantly influenced by various factors such as complexity, contrast, order, grouping, familiarity, style, and contents, among others.

In [38]'s model, the visual artifact offers stimuli, to which the users respond with attention efforts based on early vision features (such as orientation, shape, and color, etc.), and then intermediate vision (by grouping). As a result, the users make a decision influenced by emotional response.

To summarize the neural models for aesthetics, we can categorize the aesthetic characteristics derived from neural visual perception into three broad categories:

- Factors of composition, which includes elements of axis, complexity, dominance, focus, layout, order, sequence, and rules of composition;
- Principles of visual design, which includes balance, contrast, emphasis, gestalt, movement, pattern, proportion, rhythm, and unity;
- Elements of visual design, which includes color, form, line, point, shape, space, texture, typography, and value.

Therefore, the neural foundation for aesthetics can thus be modeled by Fig. 2.

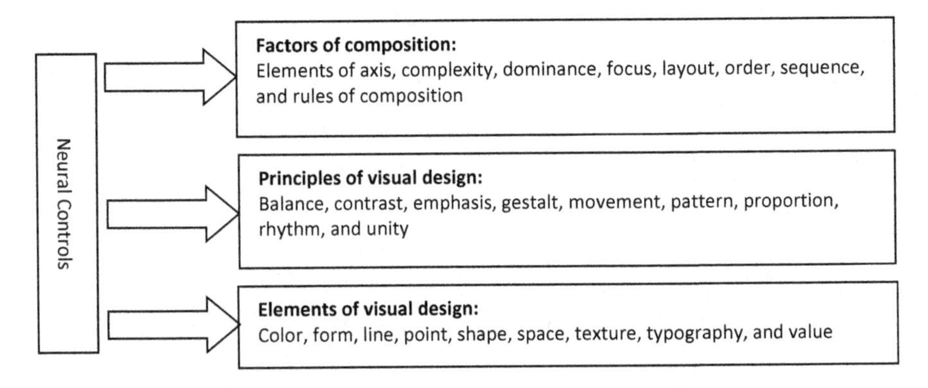

Fig. 2 The model of neural foundation for aesthetics

4 A General Framework for Neuro-Scientific IS Visual System Design

Based on the description of the previous analyses, it can be noticed that neural controls play a critical role in bridging the two seemingly disparate paradigms of IS visual system design: aesthetic design principles and positivistic design principles. Another important characteristic of this neuro-scientific integrated visual system design approach is that neural controls function in a behind-the-scene, under-the-hood fashion. Their intricate functionalities and mechanisms in influencing either positivistic IS design decisions or aesthetic design perceptions are critical in understanding how the decisions and perceptions are formed. However, these neural controls' detailed intricacies are encapsulated from the implementation of the design framework in real-world designing scenarios. People do not need to understand the detailed neural functionalities of various controls to appreciate the integration of traditional IS system design methodologies and aesthetics principles.

In Fig. 3, we show the neural factors as the bridging components in connecting aesthetics visual design dimensions (as identified by Fig. 2) and functional system design dimensions (as identified by Fig. 1).

Next, we encapsulate the neural factors underlying the connections between aesthetic dimensions and functional dimensions, and present the general framework for visual system design as illustrated by Fig. 4.

The interactive flows between all dimensions are bi-directional, meaning that the higher-level compositional requirements (Level -1 in Fig. 4) potentially dictate the knowledgeable designer's choice of the lower-level principles (Level -2), which in turn may dictate the choice of the lowest level elements (Level -3). IT system developers know that poor functional system design usually results in an inferior system. Analogously, because elementary (-3) design choices can profoundly affect principles (-2) and then exert influence through the compositional factors (-1), a designer who does not understand visual design theory and the possible hierarchical

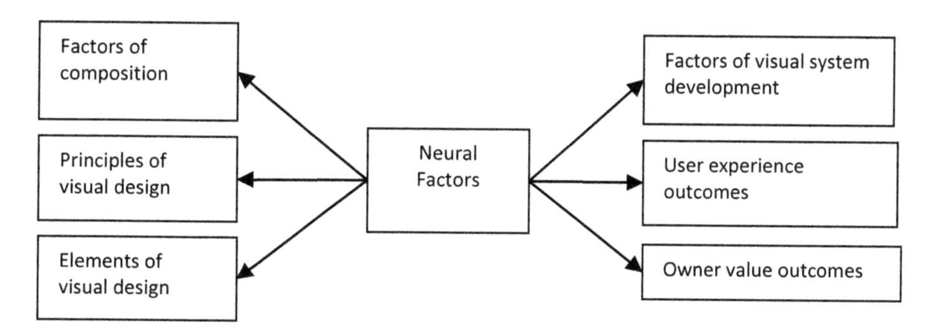

Fig. 3 Neural factors as a bridge between aesthetics and functional system design

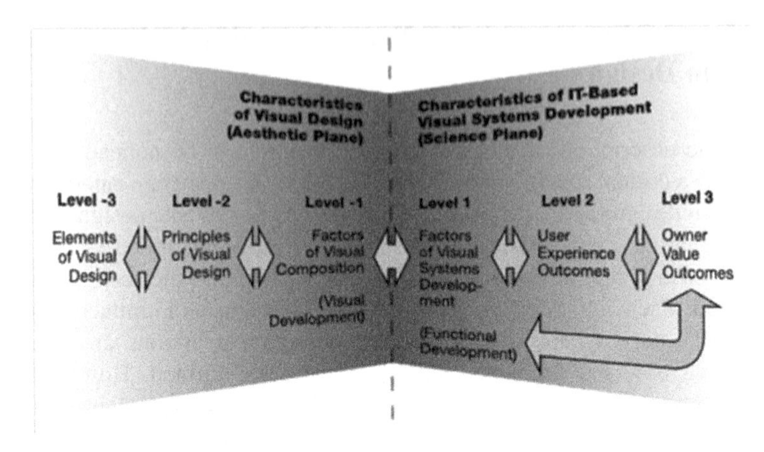

Fig. 4 A general framework for neuro-scientific visual systems design

and interactive effects of his/her design choices will likely produce inferior visual designs.

5 A Topological Framework for a Balanced Visual Systems Design

Based on our proposed general conceptual framework for neuro-scientific visual systems design, we are able to establish a typological framework of balanced visual systems design. We start with each dimension in our conceptual frame. We identify each element within the dimension as the building blocks of design, and divide each into additional qualities, or subordinate attributes, as sub-variables. Then we provide a detailed description of each sub-variable, and identify its factor type and class.

The 49 factors or variables listed in our conceptual framework collectively generate over 150 qualities. Due to the length limit of this paper, we are not able to include the whole typological framework in this paper. Figure 5 is a snapshot of the portion of the framework pertaining to the element of "color" in the dimension of "elements of visual design" within the aesthetic plane.

Paradigm	Level	Design Factor	Description	Factor Type	Class
Aesthetic			An overarching conceptual model based on theory, study, and practice of beauty	Construct	Paradigm
	-3		The first-order visual dimension of the visual aesthetic paradigmatic plane, containing compositional factors visual aesthetic design--resulting from combination and manipulation of elements and principles. All variables in this dimension are composed of sub-variable "qualities"--characteristics that help define the factors.	Dimension	Characteristic of Visual Design
		Color	The perception of particular wavelengths of light as specific hues by the color receptors of the human eye	Variable	Element of Design
			Qualities: Hue, Saturation, Luminance, Chrominance, Complementary, Associative, Analogous, Surface area occupied	*Sub-Variables*	

Fig. 5 A snapshot of the typological framework of visual systems design

6 Conclusion

In this paper, we present a general conceptual framework based on neuroscience that systematically integrates aesthetic principles and traditional positivistic IS design science methodology. In addition, based on our conceptual model, we establish a typological framework that directly provides the practical guidelines of a visual system design approach which balance the aesthetic and functional aspects of system design.

For future research, a couple of possible directions are interesting in extending the research outcome from this paper. First of all, we believe it would be interesting to design an experiment to evaluate the implementation of the typological framework developed in this paper and test how well it performs with an actual IS visual system (e.g. a functional website) design. Another interesting topic would be to obtain more details about how exactly the various neural controls affect the aesthetic perception and positivistic decision making and thus explore deeper insights towards the integrated relationships between these two planes in our proposed framework as illustrated in Fig. 4.

References

1. Pettersson, R.: Information Design: An Introduction. John Benjamins Publishing Company, Amsterdam (2002)
2. Benbasat, I.: HCI research: Future challenges and directions. AIS Trans. Hum.-Comput. Interact. **2**, 1 (2010)

3. Li, N.L., Zhang, P.: The intellectual development of human-computer interaction research: A critical assessment of the MIS literature (1990–2002). J Assoc Inform Syst **6**, 9 (2005)
4. Prestopnik, N.R.: Theory, design and evaluation–(Don't just) pick any two. AIS Trans. Hum.-Comput. Interact. **2**, 167–177 (2010)
5. Zhang, P., Benbasat, I., Carey, J., Davis, F., Galletta, D.F., Strong, D.: AMCIS 2002 panels and workshops I: Human-computer interaction research in the MIS discipline. Comm Assoc Inform Syst **9**, 20 (2002)
6. Zhang, P., Li, N., Scialdone, M., Carey, J.: The intellectual advancement of human-computer interaction research: A critical assessment of the MIS literature (1990–2008). AIS Trans. Hum.-Comput. Interact. **1**, 55–107 (2009)
7. Cyr, D., Head, M., Larios, H., Pan, B.: Exploring human images in website design: A multi-method approach. MIS Q **33**, 539–566 (2009)
8. Babu, R., Singh, R., Ganesh, J.: Understanding blind users' Web accessibility and usability problems. AIS Trans. Hum.-Comput. Interact. **2**, 73–94 (2010)
9. Palmer, J.W.: Web site usability, design, and performance metrics. Inform Syst Res **13**, 151–167 (2002)
10. Reber, R., Schwarz, N., Winkielman, P.: Processing fluency and aesthetic pleasure: Is beauty in the perceiver's processing experience? Pers Soc Psychol Rev **8**, 364–382 (2004)
11. Teo, H.-H., Oh, L.-B., Liu, C., Wei, K.-K.: An empirical study of the effects of interactivity on web user attitude. Int J Hum Comput Stud **58**, 281–305 (2003)
12. Venkatesh, V., Ramesh, V.: Web and wireless site usability: Understanding differences and modeling use. MIS Q **30**, 181–206 (2006)
13. Agarwal, R., Karahanna, E.: Time flies when you're having fun: Cognitive absorption and beliefs about information technology usage. MIS Q **24**, 665–694 (2000)
14. Beaudry, A., Pinsonneault, A.: The other side of acceptance: studying the direct and indirect effects of emotions on information technology use. MIS Q **34**, 689–710 (2010)
15. Schrepp, M., Held, T., Laugwitz, B.: The influence of hedonic quality on the attractiveness of user interfaces of business management software. Interact. Comput. **18**, 1055–1069 (2006)
16. Iivari, J., Hirschheim, R., Klein, H.K.: A paradigmatic analysis contrasting information systems development approaches and methodologies. Inform Syst Res **9**, 164–193 (1998)
17. Valacich, J.S., Parboteeah, D.V., Wells, J.D.: The online consumer's hierarchy of needs. Comm ACM **50**, 84–90 (2007)
18. Peak, D.A., Gibson, M.R., Prybutok, V.R.: Synergizing positivistic and aesthetic approaches to improve the development of interactive, visual systems design. Inform Des J **19**, 103–121 (2011)
19. Lindgaard, G., Fernandes, G., Dudek, C., Brown, J.: Attention web designers: You have 50 milliseconds to make a good first impression! Behav Inform Tech **25**, 115–126 (2006)
20. Edwards, D.J.: The Handbook of Art and Design Terms. Pearson/Prentice Hall, Upper Saddle River (2004)
21. Krug, S.: Don't Make Me Think: A Common Sense Approach to the Web, 2nd edn. New Riders, Berkeley, CA (2005)
22. Mandiberg, M.: Digital Foundations: Intro to Media Design with the Adobe Creative Suite. Peachpit Press, Berkeley, CA (2008)
23. Mullet, K., Sano, D.: Designing Visual Interfaces: Communication Oriented Techniques. Prentice Hall, Upper Saddle River, NJ (1994)
24. Rand, P.: Design Form and Chaos. Yale University Press, New Haven, CT (1993)
25. Peak, D.A., Prybutok, V.R., Gibson, M., Xu, C.: Information systems as a reference discipline for visual design. Int J Art Cult Des Tech **2**, 57–71 (2012)
26. Peak, D.A., Prybutok, V.R., Wu, Y., Xu, C.: Integrating the visual design discipline with information systems research and practice. Inf. Sci. Int. J. Emerg. Transdiscipl. **14**, 161–182 (2011)
27. Hevner, A., Chatterjee, S.: Design research in information systems: Theory and practice. Springer Science & Business Media, Heidelberg (2010)

28. Peffers, K., Tuunanen, T., Rothenberger, M.A., Chatterjee, S.: A design science research methodology for information systems research. J Manag Inform Syst **24**, 45–77 (2007)
29. Liapis, C., Chatterjee, S.: On a NeuroIS design science model. In: Service-Oriented Perspectives in Design Science Research, pp. 440–451. Springer, Berlin (2011)
30. Dimoka, A., Banker, R.D., Benbasat, I., Davis, F.D., Dennis, A.R., Gefen, D., Gupta, A., Ischebeck, A., Kenning, P.H., Pavlou, P.A., Müller-Putz, G., Riedl, R., Brocke, J.V., Weber, B.: On the use of neurophysiological tools in is research: Developing a research agenda for neurois. MIS Q **36**, 679–702 (2012)
31. Riedl, R., Banker, R.D., Benbasat, I., Davis, F.D., Dennis, A.R., Dimoka, A., Gefen, D., Gupta, A., Ischebeck, A., Kenning, P.: On the foundations of NeuroIS: Reflections on the Gmunden Retreat 2009. Comm Assoc Inform Syst **27**, 15 (2010)
32. Brocke, J.V., Riedl, R., Léger, P.-M.: Application strategies for neuroscience in information systems design science research. J Comput Inform Syst **53**, 1–13 (2013)
33. Loos, P., Riedl, R., Müller-Putz, G.R., Vom Brocke, J., Davis, F.D., Banker, R.D., Léger, P.-M.: NeuroIS: Neuroscientific approaches in the investigation and development of information systems. Bus. Inf. Syst. Eng. **2**, 395–401 (2010)
34. Riedl, R., Randolph, A.B., vom Brocke, J., Léger, P.-M., Dimoka, A.: The potential of neuroscience for human-computer interaction research. In: SIGHCI 2010 Proceedings (2010)
35. Vom Brocke, J., Riedl, R., Léger, P.-M.: Neuroscience in design-oriented research: Exploring new potentials. In: Service-Oriented Perspectives in Design Science Research, pp. 427–439. Springer, Berlin (2011)
36. Chatterjee, A.: Neuroaesthetics: A coming of age story. J Cognit Neurosci **23**, 53–62 (2011)
37. Leder, H., Belke, B., Oeberst, A., Augustin, D.: A model of aesthetic appreciation and aesthetic judgments. Br J Psychol **95**, 489–508 (2004)
38. Chatterjee, A.: Prospects for a cognitive neuroscience of visual aesthetics. Bull Psychol Arts **4**, 6 (2004)

Towards a General-Purpose Mobile Brain-Body Imaging NeuroIS Testbed

Reinhold Scherer, Stefan Feitl, Matthias Schlesinger, and Selina C. Wriessnegger

Abstract Navigating (familiar) environments requires spatial memory and spatial orientation. Mobile information systems (IS) have largely taken on this task and have changed human behavior. What impact has the redistribution of problem solving on human skills and knowledge? We are interested in exploring how the use of IS impacts on knowledge/ignorance by means of mobile brain-body imaging. In this paper, we introduce a novel experimental testbed developed to study spatial orientation in the context of geographic maps. Key system features include data synchronization between various devices and data sources, flexibility in designing and modeling research questions and integration of online co-adaptive brain-computer interfacing (BCI) technology. Flexibility, adaptability, scalability and modifiability of the implemented system turn the testbed into a general-purpose tool for studying NeuroIS constructs.

Keywords Mobile brain and body imaging • Spatial orientation • General-purpose experiment environment • Brain-computer interface

1 Introduction

Information systems (IS) and mobile devices allow accessing information and knowledge everywhere and at any time. The use of new and modern technology, hence, supports our knowledge-based society in freeing cognitive resources. People are enabled to focus on more important things. The downside of using IS is the vast amount of information available that challenges the limited capacity of our working memory. Cognitive resources are required to understand and evaluate the provided information, and to put it into the context of the current issue. Information can be blessing or curse. Accordingly, views vary on whether or not the use of IS leads to a society of knowledge or ignorance. In the former case, information is converted into new knowledge. In the latter case, information becomes a consumer product. Worst

R. Scherer (✉) • S. Feitl • M. Schlesinger • S.C. Wriessnegger
Institute of Neural Engineering, Graz University of Technology, Graz, Austria
e-mail: reinhold.scherer@tugraz.at

F.D. Davis et al. (eds.), *Information Systems and Neuroscience*, Lecture Notes in Information Systems and Organisation 16, DOI 10.1007/978-3-319-41402-7_17

case, given the fact that information can be true or false, information is blindly trusted.

Dealing with new tools and technologies is a skill that involves learning. The learning process leads to changes in human behavior, which in turn induces changes in the brain. Mobile navigation devices, for example, help finding the fastest route to a destination. Users can focus on driving. The result is safer driving. The shift of attention towards driving and interaction with the technology prevents users from the acquisition of new spatial knowledge [1]. However, not using brain areas related to spatial orientation may lead to synaptic pruning. For example, licensed London taxi drivers know the city plan inside and out and have exceptional navigation skills. These abilities are associated with an enlargement of posterior hippocampi [2, 3]. Shrinking hippocampus on the other hand signals early Alzheimer's disease [4]. One may conclude that frequent users of navigation devices (or computer games [5]) could be at increased risk of developing neurological disorders during their lifetime.

We are interested in studying neural mechanisms of decision making and application of learned understanding associated with the use of IS. In order to do this, an experimental testbed is essential that allows acquiring behavioral and neural data from people while they are engaged in tasks that require interaction with IS. This data allows for brain-body imaging [6]. What is essential is that the experimental setup is as realistic as possible and that persons are enabled to interact and behave naturally [6, 7]. Motivated by the above example of spatial navigation, our current aim is to study non-invasive electroencephalogram (EEG) correlates of spatial orientation [7, 8] in the context of geographical maps by addressing the following scientific questions: Is knowledge and education of map-reading abilities represented in EEG dynamics? Are EEG dynamics of people experienced with topographic maps different when compared to persons that are inexperienced? What is the EEG dynamics situation in persons that mainly count on mobile IS for navigation? In this paper, after reviewing the relevant literature and formulating hypotheses, we present the experimental paradigm and introduce the testbed developed to address these questions.

2 EEG Correlates of Spatial Orientation

Animal studies provided evidence that the hippocampus contains our own internal, spatial map. Particular neurons fire as function of location (place cells), distance (grid cells in the entorhinal cortex) or direction (head direction cells). Hippocampus, together with retrosplenial complex (RSC) and medial parietal cortex are brain structures that play a crucial role for spatial navigation and orientation. Functional magnetic resonance imaging (fMRI) studies confirmed that the above listed brain structures are involved in spatial navigation in humans as well [2, 3]. Activity in these structures even allows prediction of navigation performance [9–11]. Involved structures, however, are located under the cerebral cortex. It is still unclear to which extend the use of non-invasive EEG allows effective location and characterization of activity in these brain areas. The use of non-invasive mobile EEG compared to bulky and stationary fMRI technology, however, allows getting closer to reality, i.e., to natural behavior.

Our hypothesis is that people that score high on a spatial orientation test use less effort in finding coordinates on a map compared to persons that score low. Effort expenditure depends on task difficulty and motivation [12]. We believe to find differences in behavior (for example, task completion time or eye movement trajectories) and in EEG activity/connectivity patterns between the two groups. We anticipate that individuals with high scores will be able to solve the spatial orientation tasks efficiently, i.e., significantly quicker, compared to individuals with lower scores. The neural efficiency hypothesis suggests that more efficient brain functioning results in less and more focused activation [13]. This leads to the hypothesis that individuals that complete the spatial orientation tasks quickly and correctly also show reduced and more focused activity patterns compared to individuals that are not as efficient. Moreover, stronger phase locking between short-distant regions of the frontal cortex is expected in efficient performers [13].

Spatial navigation is a fundamental but complex cognitive function, which requires the integration and manipulation of multisensory information over time and space [14]. Accordingly, we can expect activity in widespread cortical networks while participants are engaged in a spatial orientation task. We hypothesize that occipital (visual information processing), parietal (sensory integration) and frontal lobes (memory and motor planning/control) show activation during the spatial orientation task when compared to a controlled rest condition. This hypothesis is based on recent findings in EEG. Motor-parietal cortical network activity was found in persons that performed an interactive gait task in a virtual environment [7, 15]. Activity in frontal cortex, motor cortex, parietal cortex, retrosplenial complex (RSC) and occipital cortex was found in persons that performed a path integration task in virtual reality [8]. In these studies, activity was reflected in task-related suppression, enhancement or modulation of theta (4–7 Hz), alpha (8–12 Hz), mu, beta (14–30 Hz) and gamma (>30 Hz) rhythms.

Spatial navigation requires a spatial representation. Spatial reference frames (SRFs) can be either allocentric (object-to-object, relative distance between objects) or egocentric (self-to-object, relative distance from self to objects) [16]. The RSC plays a central role in this context [8]. It is not clear whether SRFs are relevant for interaction with topographic maps. The RSC, however, may be active when people estimate distances between positions on a map as required in our experiment.

3 Experimental Paradigm

Theoretical and practical knowledge on geographical maps and spatial orientation skills will be assessed by a questionnaire we have developed. Individuals with highest and lowest scores are invited to participate in EEG recordings. Participants will be seated in front of a computer screen and asked to answer questions that require basic knowledge on longitude, latitude, cardinal points and the ability to estimate distances. To answer the questions participants will interact with an IS that provides interactive access to geospatial data. Possible questions include "Which city lies to the north-west of the following coordinates? 15.10 eastbound longitudes 47.40 northern latitudes?" Or "Which city is closer to the coordinates? 47.40 N, 15.10 E". Note, that different types of questioning are used so that learned knowledge has to be applied. In the first experimental condition (condition interactive), participants have the task to correctly transfer global positioning system (GPS) coordinates into the IS, cognitively process the digital map (GPS coordinates are highlighted) and answer the question by clicking on one of two possible answers. In the second experimental condition (condition static) only a static image of a map is presented. Users have to find by themselves the given GPS coordinates and answer the question. User interaction is based on keyboard and mouse. Figure 1a shows screenshots of the user interface.

User behavior is quantified by recording keyboard and mouse input, the point of eye gaze (RED/SMI Eye Tracking Glasses 2.0, SensoMotoric Instruments GmbH, Germany), muscle signals (electromyogram, EMG) from both arms and events generated by the IS that enable reconstructing the user interaction. EEG is recorded by a mobile 64-channel EEG biosignal amplifier (eegosports, ANT-Neuro, Netherlands). The next section provides technical details about the IS developed for this study.

4 Towards A General Purpose Framework

The system consists of two main components: the Lab Streaming Layer (LSL, [17]) for data synchronization, and the FrageServer for user interaction:

- LSL (available at https://code.google.com/p/labstreaminglayer) is a software system that allows collection and synchronization of data from a number of different (hardware) sources. LSL provides a network-based core transport layer as well as standalone hardware applications that utilize the layer to either make data available (e.g. data from EEG amplifier or eye-tracker) or retrieve data (e.g. data storage or online BCI processing). Available protocols and application programming interfaces allow easy expandability and integration of new sensors and devices [18].
- The FrageServer is a web-based client-server IS designed and developed to manage, plan, model and perform experiments. The platform is based on

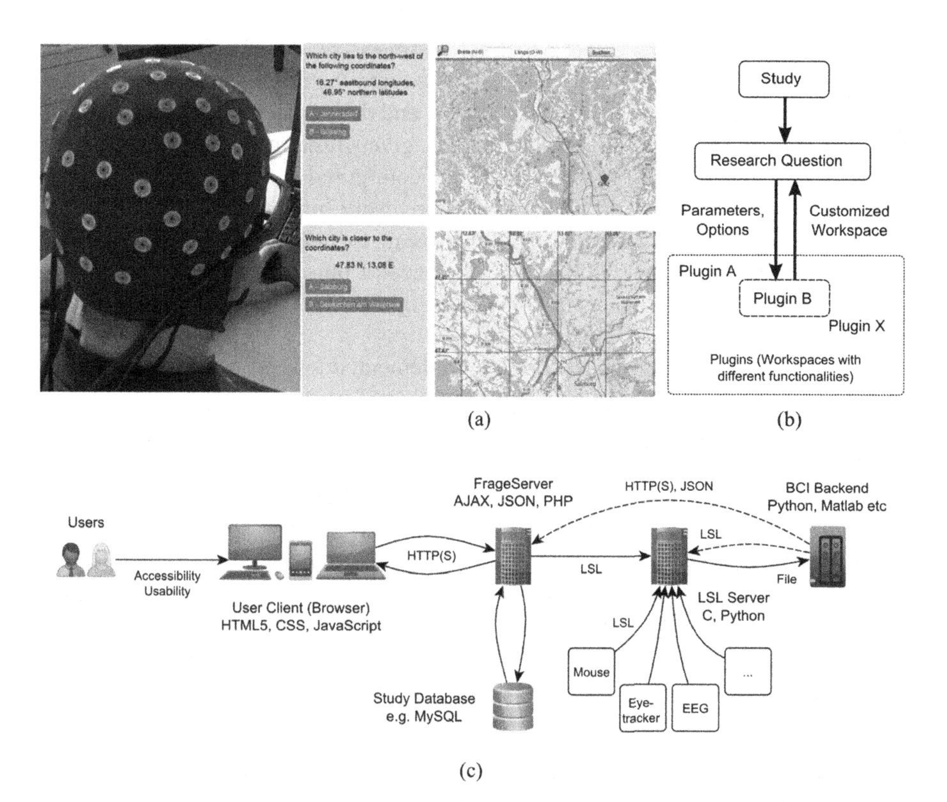

Fig. 1 (**a**) Screenshots of the user interface for interactive (*upper panel*) and static (*lower panel*) experimental conditions. Questions and possible answers are presented in the left pane. Maps are visualized in the *right pane*. The search input box for the interactive condition is located above the map. (**b**) Hierarchical representation of studies. Plugins are workspaces that provide the functionality required for each behavioral task. (**c**) Overall system architecture. Components and when applicable applied technologies and protocols are shown (Lab Streaming Layer, LSL; Brain-Computer Interface, BCI). *Arrows* illustrate the data flow between components. *Dashed lines* illustrate online processing pathways. The picture shows a person participating in a spatial orientation pilot experiment. The user is wearing a cap with EEG sensors and an eye-tracker while answering questions as asked by the experimental paradigm

state-of-the-art technologies such as HTML5, CSS3, WCAG 2.0 Level AA, AJAX, JSON, PHP, and LSL, which ensure that the implemented server runs on current web servers and clients in modern browsers. This also ensures that the same interface and functionality paired with full accessibility is available across computer systems (from PC to smartphone).

Neuroscientific studies typically aim at answering one or more (N) research questions (Q_i, $i = 1\ldots N$). In this study, our question is whether or not persons with high and low spatial orientation skills have different EEG signatures. Two or more (M) behavioral tasks (T_j, $j = 1\ldots M$) are commonly needed to answer Q_i. We designed the interactive and static spatial interaction tasks. Since behavior varies, each task T_j has to be repeated a number of times ($r = 1...R$). This hierarchical representation of study design was implemented into the FrageServer (see Fig. 1b). Plugins represent different behavioral tasks T_j and deliver a customized workspace with functionalities required for solving the given task in the front-end of the FrageServer. Additionally, each plugin can be configured with different parameters and options for every repetition r. Two plugins were implemented for the spatial orientation study: the first plug-in allows users to enter GPS coordinates and visualize OpenStreetMap (OSM) data. The second plug-in displays a coordinate grid for given GPS coordinates in OSM (see Fig. 1a). The type of questioning was defined for each repetition r individually.

The overall system architecture of the implemented testbed is shown in Fig. 1c. Users interact via the User Client with the FrageServer. Studies are stored in a database. The user interaction is streamed to the LSL server and synchronized with behavioral data (for example, keyboard, mouse, and eye-tracker) and biosignals (EEG). Any device that supports LSL can be included. The BCI backend performs online data processing. Note, that data from all LSL sources can be processed not only EEG. BCI output is sent to the FrageServer to either provide feedback to the user (for example, BCI control and communication experiments) or to modify task parameters (for example, task difficulty changes as a function of workload or affective state). Additionally, BCI output is streamed to the LSL server.

5 Discussion and Outlook

The search for neural correlates of spatial orientation led to the development of a general-purpose testbed for studying neuroscientific research questions and exploring neuro IS constructs (for example, human-computer interaction, trust/distrust, affective computing, decision making). By using state-of-the-art technologies our system runs on nearly every device with network connection. The modular architecture and the implemented plugin system allows for simple adaptation of existing and integration of new behavioral tasks. Basic plugins allow visualization of images or websites. Other workspaces such as movie player or drawing canvas can be integrated with little effort. The use of LSL allows for easy scaling of the sensor network and integration of various IS without the need of additional coding. Note, that plugins may be used to collect data from devices that do not support LSL. Since we are also working with users with functional disabilities (e.g. [19]) accessibility (e-inclusion) is important.

First pilot experiments were conducted successfully in laboratory setting with participants sitting in front of a laptop computer. Preliminary EEG analyses show meaningful results. As expected we found task-related activity over occipital, parietal and frontal cortex. However, more data has to be recorded and more in depth analyses have to be performed before specific conclusions can be made.

The selected hardware allows immediate switching from desk to mobile and ambulatory setting. Future experiments will require users to interact with mobile IS while moving freely in the environment. For example, users will have to walk to a specific destination in open space by using mobile navigation devices. Three-dimensional body movement tracking and audiovisual scene recording are needed to this end.

One issue when recording EEG while persons are in motion are bioelectrical muscle artifacts, which can produce misleading EEG signals or destroy them altogether. Recently, we made significant progress toward artifact reduction methods and the analysis of EEG while persons are in motion [15, 20–23].

We plan to conduct experiments where the tasks provided by the FrageServer change based on the user's mental state or behavioral parameters. For example, the complexity of the user interface will change based on the user's workload. Or, erroneous instructions will be given to users when they get closer (based on GPS coordinates) to the destination. These experiments allow us studying user interface design and trust issues. Since we have extensive experience with online EEG data analysis, BCI and online man-machine co-adaptation [20, 24–26] the next step is the implementation of the BCI backend functionality.

Acknowledgements The authors are thankful to Christian Kothe and David Medine (Schwartz Center for Computational Neuroscience, UC San Diego) for their support with LSL software adaptations. This work was partly funded by the Land Steiermark project MODERNE (NICHT-) WISSENSGESELLSCHAFT (grant no ABT08-21624/2014).

References

1. Ishikawa, T., Takahashi, K.: Relationships between methods for presenting information on navigation tools and users' wayfinding behavior. Cartogr. Perspect. **75**, 17–28 (2013)
2. Woollett, K., Maguire, E.A.: Acquiring 'the knowledge' of London's layout drives structural brain changes. Curr. Biol. **21**, 2109–2114 (2011)
3. Maguire, E.A., Gadian, D.G., Johnsrude, I.S., Good, C.D., Ashburner, J., Frackowiak, R.S., Frith, C.D.: Navigation-related structural change in the hippocampi of taxi drivers. Proc. Natl. Acad. Sci. U. S. A. **97**, 4398–4403 (2000)
4. Henneman, W.J.P., Vrenken, H., Barnes, J., Sluimer, I.C., Verwey, N.A., Blankenstein, M.A., Klein, M., Fox, N.C., Scheltens, P., Barkhof, F., Van Der Flier, W.M.: Baseline CSF p-tau levels independently predict progression of hippocampal atrophy in Alzheimer disease. Neurology **73**, 935–940 (2009)
5. West, G.L., Drisdelle, B.L., Konishi, K., Jackson, J., Jolicoeur, P., Bohbot, V.D.: Habitual action video game playing is associated with caudate nucleus-dependent navigational strategies. Proc. R. Soc. London B Biol. Sci. **282**, 20142952 (2015)
6. Makeig, S., Gramann, K., Jung, T.-P., Sejnowski, T.J., Poizner, H.: Linking brain, mind and behavior. Int. J. Psychophysiol. **73**, 95–100 (2009)
7. Ehinger, B.V., Fischer, P., Gert, A.L., Kaufhold, L., Weber, F., Pipa, G., König, P.: Kinesthetic and vestibular information modulate alpha activity during spatial navigation: A mobile EEG study. Front. Hum. Neurosci. **8**, 71 (2014)
8. Lin, C.-T., Chiu, T.-C., Gramann, K.: EEG correlates of spatial orientation in the human retrosplenial complex. Neuroimage **120**, 123–132 (2015)

9. Sulpizio, V., Boccia, M., Guariglia, C., Galati, G.: Functional connectivity between posterior hippocampus and retrosplenial complex predicts individual differences in navigational ability. Hippocampus (2016). doi:10.1002/hipo.22592
10. Baumann, O., Mattingley, J.B.: Medial parietal cortex encodes perceived heading direction in humans. J. Neurosci. **30**, 12897–12901 (2010)
11. Chadwick, M.J., Jolly, A.E.J., Amos, D.P., Hassabis, D., Spiers, H.J.: A goal direction signal in the human entorhinal/subicular region. Curr. Biol. **25**, 87–92 (2015)
12. Brehm, J.W., Self, E.A.: The intensity of motivation. Annu. Rev. Psychol. **40**, 109–131 (1989)
13. Neubauer, A.C., Fink, A.: Intelligence and neural efficiency: Measures of brain activation versus measures of functional connectivity in the brain. Intelligence. **37**, 223–229 (2009)
14. Wolbers, T., Hegarty, M.: What determines our navigational abilities? Trends Cogn. Sci. **14**, 138–146 (2010)
15. Wagner, J., Solis-Escalante, T., Scherer, R., Neuper, C., Müller-Putz, G.R.: It's how you get there: Walking down a virtual alley activates premotor and parietal areas. Front. Hum. Neurosci. **8**, 93 (2014)
16. Klatzky, R.L.: Allocentric and egocentric spatial representations: Definitions, distinctions, and interconnections. 1–18 (1998)
17. Delorme, A., Mullen, T., Kothe, C., Akalin Acar, Z., Bigdely-Shamlo, N., Vankov, A., Makeig, S.: EEGLAB, SIFT, NFT, BCILAB, and ERICA: New tools for advanced EEG processing. Comput. Intell. Neurosci. **2011**, 130714 (2011)
18. Schlesinger, M.: An LSL-based sensor platform for mobile brain imaging, brain-computer interfaces and rehabilitation (Master's thesis) (2016)
19. Scherer, R., Billinger, M., Wagner, J., Schwarz, A., Hettich, D.T., Bolinger, E., Lloria Garcia, M., Navarro, J., Müller-Putz, G.R.: Thought-based row-column scanning communication board for individuals with cerebral palsy. Ann. Phys. Rehabil. Med. (2015). doi:10.1016/j.rehab.2014.11.005
20. Daly, I., Scherer, R., Billinger, M., Müller-Putz, G.R.: FORCe: Fully online and automated artifact removal for brain-computer interfacing. IEEE Trans. Neural Syst. Rehabil. Eng. (2014). doi:10.1109/TNSRE.2014.2346621
21. Wagner, J., Solis-Escalante, T., Grieshofer, P., Neuper, C., Müller-Putz, G.R., Scherer, R.: Level of participation in robotic-assisted treadmill walking modulates midline sensorimotor EEG rhythms in able-bodied subjects. Neuroimage **63**, 1203–1211 (2012)
22. Scherer, R., Moitzi, G., Daly, I., Müller-Putz, G.R.: On the use of games for noninvasive EEG-based functional brain mapping. IEEE Trans. Comput. Intell. AI Games **5**, 155–163 (2013)
23. Seeber, M., Scherer, R., Wagner, J., Solis-Escalante, T., Müller-Putz, G.R.: EEG beta suppression and low gamma modulation are different elements of human upright walking. Front. Hum. Neurosci. **8**, 485 (2014)
24. Pfurtscheller, G., Müller-Putz, G.R., Schlögl, A., Graimann, B., Scherer, R., Leeb, R., Brunner, C., Keinrath, C., Lee, F.Y., Townsend, G., Vidaurre, C., Neuper, C.: 15 years of BCI research at Graz University of Technology: Current projects. IEEE Trans. Neural Syst. Rehabil. Eng. **14**, 205–210 (2006)
25. Scherer, R., Müller-Putz, G.R., Pfurtscheller, G.: Flexibility and practicality: The Graz brain-computer interface approach. Int. Rev. Neurobiol. **86**, 119–131 (2009)
26. Faller, J., Vidaurre, C., Solis-Escalante, T., Neuper, C., Scherer, R.: Autocalibration and recurrent adaptation: Towards a plug and play online ERD-BCI. IEEE Trans. Neural Syst. Rehabil. Eng. **20**, 313–319 (2012)

Differences in Reading Between Word Search and Information Relevance Decisions: Evidence from Eye-Tracking

Jacek Gwizdka

Abstract We investigated differences in reading strategies in relation to information search task goals and perceived text relevance. Our findings demonstrate that some aspects of reading when looking for a specific target word are similar to reading relevant texts to find information, while other aspects are similar to reading irrelevant texts to find information. We also show significant differences in pupil dilation on final fixations on relevant words and on relevance decisions. Our results show feasibility of using eye-tracking data to infer timing of decisions made on information search tasks in relation to the required depth of information processing and the relevance level.

Keywords Information search • Reading • Relevance • Eye-tracking • Pupillometry

1 Introduction and Related Work

Relevance assessment is one of fundamental cognitive actions performed by humans. Business Dictionary defines relevant information as "*Data which is applicable to the situation or problem at hand that can help solve a problem or contribute to a solution*" [1]. This could be in the context of just about any human task, including an ecommerce transaction. Information relevance has been conceptualized in information systems literature as a component of information quality and has been shown to contribute positively to decision-making satisfaction [2]. It also has been shown to be a critical antecedent of willingness to disclose information [3, 4]. In information retrieval systems (aka search engines) area, there is a tendency to view relevance from the system-centered perspective and consider it a binary construct that can be measured algorithmically. Operationalization of this simple view has led to the world-wide success of search engines. Yet, many scholars have been discontent with such simplified notion of relevance and over the last four decades [5] efforts in theorizing relevance as a multi-dimensional,

J. Gwizdka (✉)
School of Information, University of Texas at Austin, Austin 78701, TX, USA
e-mail: neurois2016@gwizdka.com

F.D. Davis et al. (eds.), *Information Systems and Neuroscience*, Lecture Notes in Information Systems and Organisation 16, DOI 10.1007/978-3-319-41402-7_18

multi-stage and multi-valued concept continued [6–10]. We contribute to these efforts by bringing neuro-physiological methods to investigating cognitive aspects of relevance. We have previously reported our earlier studies, including an fMRI study that examined differences in brain activations between reading relevant and irrelevant texts [11] and an eye-tracking study, in which we showed measures correlated with processing of relevant and irrelevant texts [12, 13]. In this paper we present a new analysis of eye-tracking data with a focus on reading strategies on tasks with word search targets which we contrast with strategies on tasks with information search goals. The presented analysis uses pupil data in addition to eye fixation measures.

Related Work We selectively review only a few works that employed eye-tracking in investigating information relevance. Ajanki et al. [14] used eye-movement based features as implicit relevance feedback. They found regressions and relative duration of the first fixation on a word to be important features and demonstrated that information retrieval system's performance could be improved modestly by using eye-based features to select additional query terms. Not relevant documents were shown to impose lower mental load than relevant ones [12]. Buscher et al. [15] demonstrated a relationship between several eye movement measures and text passage relevance. They found the length of coherently read text to be the strongest relevance indicator. Fixation duration was not an effective discriminator between relevant and not relevant text. The reading eye movement measures were used to improve relevance ranking by ~8 % overall. For queries with poor search results, the improvement was 27 %. Simola et al. [16] used eye movement features to improve classification of processing states on three simulated information search tasks (word search, question-answer, and subjective interest) to 60.2 %. Their work used simple search tasks and single sentences rather than text paragraphs.

Most eye-trackers measure pupil dilation and this variable is also of interest to our work. Pupil dilation is controlled by the Autonomic Nervous System (ANS) [17]. Under constant illumination it has been associated with a number of cognitive functions, including mental workload [18], interest [19], surprise [20], and making decisions [21]. It is reasonable to expect that document relevance will affect level of attention or mental workload and, thus, pupil size. Only few published works examined pupil size in relation to information relevance. In [22] Oliveira et al. show reported that pupil dilated for higher relevance stimuli for text documents and images. In our earlier work on relevance of short text documents [12] and web pages [23] we showed significant pupil dilation on relevant documents and, in particular, in the one or two second period preceding relevance decision.

In this work we extend previous results by investigating how differences in reading documents that result from user tasks (locating a word and finding required information) and varying information relevance are reflected in measures obtained from eye-tracking. Our research questions are as follows: *RQ1*. Do people read differently when locating a target word as compared to when finding required information? *RQ2*. Does pupil dilation differ between the two cases?

2 Method

We conducted a controlled lab experiment, in which participants (N = 24; 9 females) were asked to find information in short text documents containing news stories. Participants were recruited from undergraduate and graduate student body at Rutgers University in New Jersey, USA. The experiment was conducted in a usability lab equipped with Tobii T-60 remote eye-tracker. The within-subjects experimental design (shown in Fig. 1) is essentially the same as reported at NeuroIS'2014 [13]; we describe it briefly below. Each participant performed two types of tasks, word search (W task), and a question-prompted information search (I task). In both tasks search was performed in short news stories (mean length 178 (30) words) selected from a large set obtained from the AQUAINT corpus [24]. Each session included 21 pseudo-randomized trials of each task type, as well as a few training trials. In W task, a trial comprised a target word presentation followed by a text that contained the target word. Each document contained exactly one target word; locating it required orthographic matching without the need for semantic processing [25]. Participants responded by pressing a key as soon as they located the target word. In I task, a trial consisted of question that instructed participants what information they were expected to find in one or more of the subsequently presented three news stories of varied relevance: *irrelevant* (*I*), *partially relevant* texts that were on a question's topic, but did not contain the answer (*T*), and *relevant* texts that contained the answer (*R*). Participants responded by judging document relevance of 63 news stories on a binary scale (yes/no). In contrast to W task, making relevance decisions in I task required semantic processing. The order of trials in both tasks and the order of text relevance in I task, were randomized. We previously reported [12, 13] earlier results from eye-tracking data analysis for I task only; in this paper we present new analysis that compares task W with I.

Dependent variables derived from eye-tracking data included two groups: (1) eye fixation variables obtained by modeling reading as a two-state process (RM-Eye variables), and (2) relative pupil dilation (RPD). Our approach to modeling reading is influenced by EZ-Reader model [26] and we have previously reported its details [13, 27]. Our reading model is line-oriented and consists of two states, (a) reading state represents continuous reading in one line; reading in a subsequent line is represented by a new reading state; (b) scanning state represents isolated fixations. Fixations are labeled as belonging to *reading* or *scanning*; before labeling, non-lexical fixations (<150 ms) are removed. Variables obtained from the reading model included probabilities of remaining in a scanning (pSS) or reading state (pRR), counts and durations of reading and scanning states and fixations. The controlled lab environment, consistent use of text documents with black background and a similar number of words presented in white font ensured virtually no variability of luminance across all trials and thus no resulting changes in pupil dilation. To eliminate individual variability in pupil sizes, we calculated a baseline for each participant (i) by taking an average pupil size over all text document

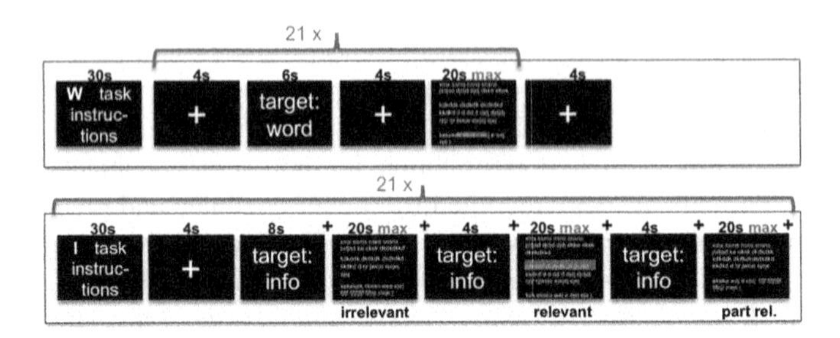

Fig. 1 Experimental design. The '+' indicates display of a centered white fixation cross

presentations ($P^{i}_{baseline}$) and calculating relative change in pupil dilation (RPD^{i}_{t}) from pupil measurement at a time t P_t as shown in Eq. (1).

$$RPD^{i}_{t} = \frac{(P_t - P^{i}_{baseline})}{P^{i}_{baseline}} \tag{1}$$

Mean values of *RPD* were calculated on each trial in three one-second-long epochs: (1) epochs starting at the first fixation on a relevant word, (2) the last fixation on a relevant word; and (3) epochs at the end of trials, that is, when participants were making relevance decisions. To avoid a possible influence of motor actions, the last 200 ms of a trial was not included.

Independent variables included, task type (W or I) and a participant's relevance judgment (*perceived relevance: Irrelevant or Relevant*).

Eye-tracking data was cleaned by removing bad quality fixations and those outside screen. This process resulted in removal of ~5 % of fixations. Raw pupil data was de-noised using Daubechies D4 level 4 wavelet transform, then algorithm adapted from [28] was used to detect blinks and replace missing data with cubic splines.

3 Data Analysis and Results

The overall accuracy of participants' relevance judgments was quite high at 90.3 % and we concluded that participants performed it according to our expectation. In our analysis, we only considered data from trials with correct responses. Due to not-normal distribution of RM-Eye variables, we analyzed them using non-parametric Kruskal-Wallis test. A three level factor was constructed by combining two-level of task types with two-levels of document relevance judgment (applicable only to I task) (Table 1). Relative pupil dilation was distributed normally, and thus we analyzed the effects of tasks and the three-level factor using one-way ANOVA (Table 2).

Table 1 Differences in RM-eye variables between tasks and texts (mean (SD))

	W Task	I Task–R	I Task–I	K-W $\chi^2(2)$
Time to complete [s]	8.4 (4.4)	9.7 (4.2)	9.6 (4.6)	17.26****
pSS	0.42 (0.25)	0.37 (0.27)	0.50 (0.26)	39.27****
pRR	0.87 (0.11)	0.89 (0.09)	0.81 (0.12)	114.38****
Fixation count on words	26 (16)	32 (17)	31 (17)	20.65****
Count of reading fixations	21.8 (14.2)	26.9 (15.7)	23.8 (16.2)	17.86****
Count of scanning fixations	5.8 (4.7)	5.3 (3.9)	8.3 (5.1)	105.13****
Total reading duration [s]	5.4 (3.4)	6.6 (3.7)	5.8 (3.9)	18.96****
Total scanning duration [s]	1.6 (1.4)	1.4 (1.0)	2.3 (1.4)	104.67****

*$p < 0.05$, **$p < 0.01$, ***$p < 0.001$, ****$p < 0.0001$

Table 2 Differences in relative pupil dilation between tasks (mean (SD)) [%]

Epoch (1s)	W Task	I Task	ANOVA
Relevance decision	0.63 (5.5)	1.02 (5.4)	$F(1,980) = 1.09$

4 Discussion and Conclusions

Our results demonstrate differences in reading when locating a target word (W task) as compared to when finding required information (I task) (*RQ1*). Overall, on W task participants were faster and had fewer fixations than on I task. Time spent scanning on W task was about the same as scanning time on relevant texts on I task and lower than scanning time on irrelevant texts on I task. In contrast, time spent reading on W task was similar to reading time on irrelevant texts on I task and lower than reading time on relevant texts on I task. At the same time, the tendency to keep reading on W task was about the same as on relevant texts on I task and both were higher than on irrelevant texts in I task. The tendency to keep scanning was highest on irrelevant documents on I task, as was the number of scanning fixations. These results show a clearly different reading pattern for each of the investigated conditions.

Even more interesting are our findings on pupil dilation (*RQ2*). We found no difference in pupil dilation between the two tasks during the relevance decision epoch (Table 2). However, we found significant differences when considering text relevance on I task (Table 3). Pupil dilation during relevance decisions on relevant texts on I task was significantly higher (3.09 %) than on W task (0.63 %) and on irrelevant texts on I task (−0.08 %). Furthermore, pupil dilation during the last fixations on relevant words was significantly higher on relevant texts on I task (2.82 %) than on W task (1.05 %). These results indicate, plausibly, that relevance decisions that involve semantic processing (I task) are of different nature than decisions that involve orthographic processing (W task) and that this difference can be inferred from changes in pupil dilation.

Practical implication of our results is that they indicate a feasibility of using eye-tracking data to infer timing of decisions made on search tasks in relation to the

Table 3 Differences in relative pupil dilation for three level factor (mean (SD)) [%]

Epoch (1s)	W Task	I Task–R	I Task–I	ANOVA
First relevant word	0.35 (5.4)	0.86 (5.1)	n/a	$F(1,393) = 0.92$
Last relevant word	1.05 (5.6)	2.82 (5.0)	n/a	$F(1,393)^a = 11.05^{***}$
Relevance decision	0.63 (5.5)	3.09 (4.9)	−0.08 (5.3)	$F(2,979)^b = 27.41^{****}$

[a]Due to lack of homogeneity of variance for this variable, we report Brown-Forsythe robust test of equality of means

[b]Post hoc analysis (Games-Howell) indicated a significant difference between I task–R and other conditions; there was no significant difference between W task and I task–I. Statistical significance is marked by: ***$p < 0.001$, ****$p < 0.0001$

required depth of information processing and the relevance level. Analysis presented in this paper was at the aggregate level of all participants. In future work, we plan to perform analysis for individual participant. We also plan to conduct more detailed analysis of pupillary responses.

Acknowledgements This research has been supported, in part, by IMLS Career Award #RE-04-11-0062-11 and by Google Faculty Research Award to Jacek Gwizdka.

References

1. Business Dictionary: Relevant information. http://www.businessdictionary.com/definition/relevant-information.html
2. Bharati, P., Chaudhury, A.: An empirical investigation of decision-making satisfaction in web-based decision support systems. Decis. Support Syst. **37**, 187–197 (2004)
3. Zimmer, J.C., Arsal, R.E., Al-Marzouq, M., Grover, V.: Investigating online information disclosure: Effects of information relevance, trust and risk. Inf. Manage. **47**, 115–123 (2010)
4. Li, H., Sarathy, R., Xu, H.: The role of affect and cognition on online consumers' decision to disclose personal information to unfamiliar online vendors. Decis. Support Syst. **51**, 434–445 (2011)
5. Saracevic, T.: RELEVANCE: A review of and a framework for the thinking on the notion in information science. J. Am. Soc. Inf. Sci. **26**, 321–343 (1975)
6. Huang, X., Soergel, D.: Relevance: An improved framework for explicating the notion. J. Am. Soc. Inf. Sci. Technol. **64**, 18–35 (2013)
7. Borlund, P.: The concept of relevance in IR. J. Am. Soc. Inf. Sci. Technol. **54**, 913–925 (2003)
8. Zhang, Y., Zhang, J., Lease, M., Gwizdka, J.: Multidimensional relevance modeling via psychometrics and crowdsourcing. In: Proceedings of the 37th International ACM SIGIR Conference on Research and Development in Information Retrieval, pp. 435–444. ACM, New York (2014)
9. da Costa Pereira, C., Dragoni, M., Pasi, G.: Multidimensional relevance: Prioritized aggregation in a personalized information retrieval setting. Inf. Process. Manag. **48**, 340–357 (2012)
10. Cosijn, E., Ingwersen, P.: Dimensions of relevance. Inf. Process. Manag. **36**, 533–550 (2000)
11. Gwizdka, J.: Looking for information relevance in the brain. In: Gmunden Retreat on NeuroIS 2013, p. 14. Gmunden, Austria (2013)
12. Gwizdka, J.: Characterizing relevance with eye-tracking measures. In: Proceedings of the 5th Information Interaction in Context Symposium, pp. 58–67. ACM, New York (2014)

13. Gwizdka, J.: Tracking information relevance. In: Gmunden Retreat on NeuroIS 2014, p. 3. Gmunden, Austria (2014)
14. Ajanki, A., Hardoon, D., Kaski, S., Puolamäki, K., Shawe-Taylor, J.: Can eyes reveal interest? Implicit queries from gaze patterns. User Model. User-Adapt. Interact. **19**, 307–339 (2009)
15. Buscher, G., Dengel, A., Biedert, R., Elst, L.V.: Attentive documents: Eye tracking as implicit feedback for information retrieval and beyond. ACM Trans. Interact. Intell. Syst. **1**, 9:1–9:30 (2012)
16. Simola, J., Salojärvi, J., Kojo, I.: Using hidden Markov model to uncover processing states from eye movements in information search tasks. Cogn. Syst. Res. **9**, 237–251 (2008)
17. Onorati, F., Barbieri, R., Mauri, M., Russo, V., Mainardi, L.: Characterization of affective states by pupillary dynamics and autonomic correlates. Front. Neuroengineering. **6**, 9 (2013)
18. Kahneman, D., Beatty, J.: Pupil diameter and load on memory. Science **154**, 1583–1585 (1966)
19. Krugman, H.E.: Some applications of pupil measurement. J. Mark. Res. (pre-1986) **1**, 15 (1964)
20. Preuschoff, K., 't Hart, B.M., Einhäuser, W.: Pupil dilation signals surprise: Evidence for noradrenaline's role in decision making. Front. Decis. Neurosci. **5**, 115 (2011)
21. Einhäuser, W., Koch, C., Carter, O.L.: Pupil dilation betrays the timing of decisions. Front. Hum. Neurosci. **4**, 18 (2010)
22. Oliveira, F.T.P., Aula, A., Russell, D.M.: Discriminating the relevance of web search results with measures of pupil size. In: Proceedings of the 27th International Conference on Human Factors in Computing Systems, pp. 2209–2212. ACM, Boston (2009)
23. Gwizdka, J., Zhang, Y.: Differences in eye-tracking measures between visits and revisits to relevant and irrelevant web pages. In: Proceedings of the 38th International ACM SIGIR Conference on Research and Development in Information Retrieval, pp. 811–814. ACM, New York (2015)
24. Graff, D.: The AQUAINT Corpus of English News Text. (2002). Available at: http://www.language-archives.org/item/oai:www.ldc.upenn.edu:LDC2002T31
25. Léger, L., Rouet, J.-F., Ros, C., Vibert, N.: Orthographic versus semantic matching in visual search for words within lists. Can. J. Exp. Psychol. Rev. Can. Psychol. Expérimentale. **66**, 32–43 (2012)
26. Reichle, E.D., Pollatsek, A., Rayner, K.: EZ reader: A cognitive-control, serial-attention model of eye-movement behavior during reading. Models Eye-Mov. Control Read. **7**, 4–22 (2006)
27. Cole, M.J., Gwizdka, J., Liu, C., Bierig, R., Belkin, N.J., Zhang, X.: Task and user effects on reading patterns in information search. Interact. Comput. **23**, 346–362 (2011)
28. Mathôt, S.: A simple way to reconstruct pupil size during eye blinks. 10.6084/m9.figshare.688001 (2013)

The Influence of Task Characteristics on Multiple Objective and Subjective Cognitive Load Measures

Seyed Mohammad Mahdi Mirhoseini, Pierre-Majorique Léger, and Sylvain Sénécal

Abstract Using Electroencephalography (EEG), this study aims at extracting three features from instantaneous mental workload measure and link them to different aspect of the workload construct. An experiment was designed to investigate the effect of two workload inductors (Task *difficulty* and *uncertainty*) on extracted features along with a subjective measure of mental workload. Results suggest that both subjective and objective measures of workload are able to capture the effect of task *difficulty*; however only accumulated load was found to be sensitive to task *uncertainty*. We discuss that the three EEG measures derived from instantaneous workload can be used as criteria for designing more efficient information systems.

Keywords NeuroIS • Mental workload • Accumulated load • Instantaneous load • Peak load • Electroencephalography

1 Introduction

Measuring the mental workload construct has been a challenge to researchers in different fields [1–3]. This construct has either been assessed subjectively with self-reported measures or objectively with measures such as electroencephalography (EEG). Subjective workload measures prevent us from understanding the multiple variations of workload during a task and are subject to a retrospective bias. The use of neurophysiological workload measures can alleviate these shortcomings to some extent [4]. Researchers have compared subjective and objective mental workload and suggested that objective measures can provide a more comprehensive and richer understanding of the workload construct [5, 6].

However, thus far, research has only used one of many possible measures of objective workload when comparing it to subjective workload [7]. Given that multiple measures of objective workload can be extracted from EEG signals (e.g., *Average load, Accumulated load, and Number of peaks*), an even richer

S.M.M. Mirhoseini (✉) • P.-M. Léger • S. Sénécal
HEC Montreal, Montreal, QC, Canada
e-mail: seyedmohammadmahdi.mirhoseini@hec.ca

F.D. Davis et al. (eds.), *Information Systems and Neuroscience*, Lecture Notes in Information Systems and Organisation 16, DOI 10.1007/978-3-319-41402-7_19

understanding of cognitive load could be attained. Thus, the objective of this research is to investigate the relationships between multiple objective workload measures and a subjective workload measure. In addition, it aims at better understanding the effects of task characteristics (i.e., uncertainty and difficulty) on these different workload measures.

2 Literature Review

Researchers from various disciplines have been interested in studying cognitive workload [8, 9], such as understanding the consequences of various workload levels on user performance [10]. Mental workload can be defined as "the set of mental resources that people use to encode, activate, store, and manipulate information while they perform a cognitive task" [11]. Researchers study cognitive load in order to prevent people from experiencing high mental workload and its negative consequences, such as frustration, negative affect, or mental fatigue [12]. Lower cognitive load has also been related to users' satisfaction with online task [13].

Generally, there are three types of cognitive load measures: subjective, performance, and physiological [14, 15]. Subjective measures, which are the most mature ones, have been used for many years to study users' behavior [16]. Borrowing from neuroscience and neuropsychology, researchers in social science have started to use physiological metrics to measure cognitive load [13]. Researchers have developed several algorithms based on EEG signals to calculate mental workload metrics [17]. EEG signals are high dimensional noisy time series, which encompass a high volume of information [18]. In order to relate such signals to specific mental states (e.g., mental workload), first of all noise should be removed from the signal, and more importantly, relevant signal features should be selected that represent the desired mental state [19]. In this study, we use education literature to define three features of instantaneous workload measure.

Xie and Salvendy [7] defined four types of workload: peak load, average load, accumulated load, and overall load. Instantaneous load shows the dynamics of cognitive load over time. Peak load is the maximum load an individual experiences during a task, while accumulated load is the summation of all the cognitive load. Average load reflects the mean load an individual experience in performing a task, and overall load is the self-perceived load [7]. Figure 1 illustrates how peak load, accumulated load, and average load are related to instantaneous load. Despite the fact that all workload measures represent users' cognitive load level, they are different. Average load is an overall measure which represents mostly the amount of cognitive resources that a task requires on average. Accumulated load provides the whole amount of load that is experienced taking into account the average and the time dimensions. Therefore, it is a more comprehensive measure of workload. Another type of information that cannot be obtained using a subjective measure, average, or accumulated load is the maximum load. We know from the literature that our cognitive capacity is limited [10] and it has a red line for information

Fig. 1 Workload features

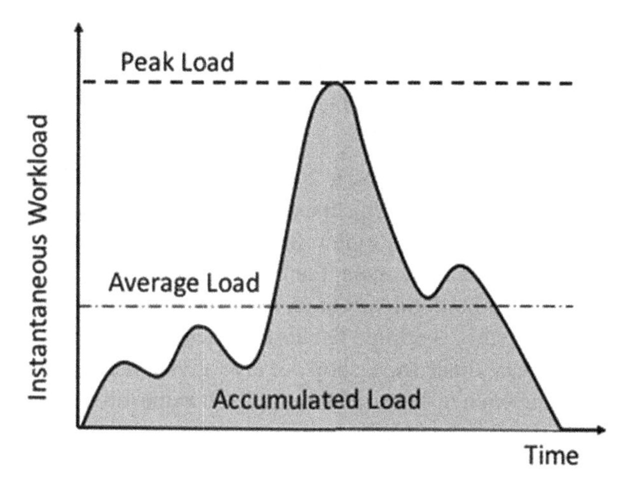

processing [20]. Other workload measures do not indicate if users have experienced such red lines. This information can be useful for studying users' coping behavior with high workload situations. For instance, if users experience high cognitive load during a difficult task they might withdraw from the task [21]. Peak load may be the proper criteria to study such behaviors.

3 Hypotheses

Tasks characteristics are an important factor in predicting users' cognitive load. In general, any factor that renders decision making more difficult will impose more load on cognitive resources. For instance, research suggests that mathematical complexity and arithmetic operations increase users' mental workload [9]. Therefore, task difficulty will affect users' mental workload.

H1 Task difficulty is positively associated with all workload types (Overall, Average, Accumulated, Peak).

Information processing includes three stages: information perception, decision/ response selection, and response execution [7]. Any factor that impairs this process may force individuals to use more cognitive resources in order to fulfill decision requirements. Information uncertainty affects the first stage by making it more difficult for users to assess the critical information required for making product decisions, which consequently imposes more cognitive load on users. A number of studies have found evidence of such effect [22].

H2 Task uncertainty is positively associated with all workload types (Overall, Average, Accumulated, Peak).

4 Methodology

4.1 Experimental Design

To test our hypotheses, a 2 (low or high task *difficulty*) × 2 (low or high task *uncertainty*) within-subject experiment was used. This experiment got the approval by our Institutional Review Board (IRB). Ten subjects participated in the experiment and 50 % were male. Each subject performed all four tasks (one per condition) which were randomly ordered. For each task, participants had to shop on a selected online grocery website. The task started on an online grocery recipe page. Participants were instructed shop for five given items for each assigned recipe. After finishing each task, subjects completed a questionnaire. Online grocery is a suitable context for this study because it is a long user-IT transaction [23]. As opposed to other online shopping contexts which are mostly composed of singles product decisions. Online grocery usually includes several decisions and allows us to observe the dynamics of users' cognitive load. Moreover, there are a number of workload antecedents which are unique to this context such as performing simple arithmetic operations in order to find the right quantity of a product or the uncertainty involved in buying perishable products.

Task *difficulty* was manipulated by asking subjects to change the quantity of ingredients suggested for the recipe. Recipe pages suggested the right amount of each grocery item to use for four people. In the low *task difficulty* condition, participants had to choose the same quantity and in the high *difficulty* condition, they were asked to perform simple arithmetic operations in order to find the right quantity of each ingredient for 20 people. We used product type (perishable or non-perishable) to manipulate task *uncertainty*. In the low *uncertainty* condition, participants had to shop for five specified non-perishable products (e.g., olive oil) and in the high *uncertainty* condition, they were instructed to shop for five specified perishable products (e.g., meat).

4.2 Measures

Overall load was measured using a five item scale developed by Cameron [24]. Instantaneous measure of cognitive load was measured using a linear EEG algorithm. In general, EEG oscillations are categorized into four frequency bands: Delta (0 to <4 Hz), Theta (4 to <8 Hz), Alpha (8 to 13 Hz), and beta (>13 Hz) [25]. In this experiment, we use a linear algorithm to measure the cognitive load of users. This algorithm includes calculating the ((delta + theta)/alpha) power ratio over a moving 2 s window and compare it with the average of previous 20 s [17]. The average of users' workload over a task period was used as *Average load*. The area under the instantaneous workload curve which equals the sum of instantaneous load over time was used as *Accumulated load*. To calculate the *Peak load*, we counted the number of times that the amplitude of instantaneous load exceeded 2.5 standard deviations of the instantaneous load, indicating the number of times participants experienced nearly peak load.

5 Apparatus, Data Acquisition, and Analysis

EEG data was recorded using a 32 electrodes using EGI's dense array electroencephalography (dEEG). In order to test the hypotheses, we used a regression analysis with task *difficulty* and *uncertainty* as independent variables and mental workload (*Overall, Average, Accumulated, Peak loads*) as dependent variable. Since the observations are non-independent, we used regression analysis for repeated measures. We also controlled for learning effect by including the order in which participants performed the task into the model.

6 Results

H1 proposed that task *difficulty* is positively associated with all measures of mental workload. Results show significant relationships between *difficulty* and Overall load ($b = 0.88$, $p < 0.01$), Average load ($b = 0.02$, $p < 0.05$), Accumulated load ($b = 44.45$, $p < 0.01$), number of Peaks ($b = 4.87$, $p < 0.05$). Therefore, H1 is supported. Our second hypothesis proposed that task *uncertainty* is an antecedent of all workload types. Regression results show that the relationships between *uncertainty* and Overall load ($b = 0.51$, $p = 0.18$), Average load ($b = 0.01$, $p = 0.15$), number of Peaks ($b = 3.12$, $p = 0.15$) are not significant. However, a significant result was found for the relationship between *uncertainty* and Accumulated load ($b = 35.24$, $p < 0.05$). Thus, H2 is partially supported.

7 Discussion

Our results suggest that all the extracted features of instantaneous load and the subjective measure of workload (i.e., Overall load) are sensitive to task *difficulty*; however only Accumulated load was able to capture the mental workload induced by task *uncertainty*. Based on the definition of Accumulated load, we can expect more comprehensiveness of this measure compared to Average load or Overall load. Accumulated load measures the change of users' instantaneous workload over time, thus accounting not only for overall load but also the total time that user has been performing the task. "Time on task" has been used as a measure for workload before [26], thus it may represent more dimensions of the workload construct.

To better understand the difference between workload measures, we conducted a post hoc analysis and compared the effect of task *difficulty* and *uncertainty* on each workload measure. We used four regression models with the same independent variables (Task *difficulty* and *uncertainty*) but with four different measures of workload as dependent variable (Overall, Accumulated, Average, and number of Peaks). Since the number of independent variables are the same, we can compared R-square across the regression models. Accumulated load explained more variance than other models (21 %). It was followed by number of Peaks (16 %), Overall load (14 %), and Average load (5 %). These results confirm our expectations that Accumulated load can capture more variability caused by workload inductors. Results also show a relatively high R-square for number of Peaks. In addition to being used as a workload measure, Peak load can also to identify the moments that users experience high workload. In other words, it is able to identify the exact time periods that users approach or pass the red line of their cognitive capacity. This type of information, which can only be gained using Peak load, may provide a proper lens to study users coping behavior in dealing with high workload situations.

8 Conclusion

Our analysis showed that Average load and overall load yield similar results while Accumulated load is a stronger indicator of the total workload experienced by users. Number of Peaks is also an appropriate metric to assess users' mental workload and study the consequences of experiencing high workload. Our results show that only Accumulated load was able to capture the effect of task *uncertainty*. Finally, Peak load can be used to analyze users' dynamic cognition. This research contributes to the literature by (1) introducing a mental workload feature extraction method in order to benefit from the richness of EEG data; (2) deriving three new metrics for measuring mental workload and explaining how these measures can be useful to study users' experience.

References

1. Gopher, D., Braune, R.: On the psychophysics of workload: Why bother with subjective measures? Hum. Factors J. Hum. Factors Ergon. Soc. **26**, 519–532 (1984)
2. Paas, F., Sweller, J.: An evolutionary upgrade of cognitive load theory: Using the human motor system and collaboration to support the learning of complex cognitive tasks. Educ. Psychol. Rev. **24**, 27–45 (2011)
3. Paas, F., Tuovinen, J.E., Tabbers, H., Van Gerven, P.W.: Cognitive load measurement as a means to advance cognitive load theory. Educ. Psychol. **38**, 63–71 (2003)
4. Riedl, R., Léger, P.-M.: Tools in NeuroIS Research: An Overview. In: Fundamentals of NeuroIS, pp. 47–72. Springer, Berlin (2016)

5. Ortiz De Guinea, A., Titah, R., Léger, P.-M.: Measure for measure: A two study multi-trait multi-method investigation of construct validity in IS research. Comput. Hum. Behav. **29**, 833–844 (2013)
6. de Guinea, A.O., Titah, R., Léger, P.-M.: Explicit and implicit antecedents of users' behavioral beliefs in information systems: A neuropsychological investigation. J. Manag. Inf. Syst. **30**, 179–210 (2014)
7. Xie, B., Salvendy, G.: Prediction of mental workload in single and multiple tasks environments. Int. J. Cogn. Ergon. **4**, 213–242 (2000)
8. De Jong, T.: Cognitive load theory, educational research, and instructional design: Some food for thought. Instr. Sci. **38**, 105–134 (2010)
9. Ryu, K., Myung, R.: Evaluation of mental workload with a combined measure based on physiological indices during a dual task of tracking and mental arithmetic. Int. J. Ind. Ergon. **35**, 991–1009 (2005)
10. Wickens, C.D.: Multiple resources and performance prediction. Theor. Issues Ergon. Sci. **3**, 159–177 (2002)
11. DeStefano, D., LeFevre, J.-A.: Cognitive load in hypertext reading: A review. Comput. Hum. Behav. **23**, 1616–1641 (2007)
12. Mizuno, K., Tanaka, M., Yamaguti, K., Kajimoto, O., Kuratsune, H., Watanabe, Y.: Mental fatigue caused by prolonged cognitive load associated with sympathetic hyperactivity. Behav. Brain Funct. **7**, 1 (2011)
13. Gwizdka, J.: Distribution of cognitive load in web search. J. Am. Soc. Inf. Sci. Technol. **61**, 2167–2187 (2010)
14. Gopher, D., Donchin, E.: Workload: An Examination of the Concept. In: Boff, K.R., Kaufman, L., Thomas, J.P. (eds.) Handbook of Perception and Human Performance, Vol II, Cognitive Processes and Performance, pp. 41:1–41:49. Wiley, New York (1986)
15. O'Donnell, R.D., Eggemeier, F.T.: Workload Assessment Methodology. In: Boff, K., Kaufman, L., Thomas, J. (eds.) Handbook of Perception and Human Performance, Vol II, Cognitive Processes and Performance. Wiley, New York (1986)
16. Colle, H.A., Reid, G.B.: Double trade-off curves with different cognitive processing combinations: Testing the cancellation axiom of mental workload measurement theory. Hum. Factors J. Hum. Factors Ergon. Soc. **41**, 35–50 (1999)
17. Coyne, J.T., Baldwin, C., Cole, A., Sibley, C., Roberts, D.M.: Applying Real Time Physiological Measures of Cognitive Load to improve training. In: Foundations of Augmented Cognition. Neuroergonomics and Operational Neuroscience, pp. 469–478. Springer, Berlin (2009)
18. Garrett, D., Peterson, D.A., Anderson, C.W., Thaut, M.H.: Comparison of linear, nonlinear, and feature selection methods for EEG signal classification. Neural Syst. Rehabil. Eng. IEEE Trans. On. **11**, 141–144 (2003)
19. Brouwer, A.-M., Zander, T.O., van Erp, J.B., Korteling, J.E., Bronkhorst, A.W.: Using neurophysiological signals that reflect cognitive or affective state: six recommendations to avoid common pitfalls. Front. Neurosci. **9**, 136 (2015)
20. Colle, H.A., Reid, G.B.: Estimating a mental workload redline in a simulated air-to-ground combat mission. Int. J. Aviat. Psychol. **15**, 303–319 (2005)
21. Venables, L., Fairclough, S.H.: The influence of performance feedback on goal-setting and mental effort regulation. Motiv. Emot. **33**, 63–74 (2009)
22. Aljukhadar, M., Senecal, S., Daoust, C.-E.: Using recommendation agents to cope with information overload. Int. J. Electron. Commer. **17**, 41–70 (2012)
23. Desrocher, C., Léger, P.-M., Sénécal, S., Pagé, S.-A., Mirhoseini, S.: The influence of product type, mathematical complexity, and visual attention on the attitude toward the website: The case of online grocery shopping. Presented at the Fourteenth Pre-ICIS SIG-HCI Workshop, Fort Worth, TX, Dec 2015

24. Cameron, A.-F.: Juggling Multiple Conversations with Communication Technology: Towards a Theory of Multi-communicating Impacts in the Workplace. Queen's University, Kingston, ON (2007)
25. Libenson, M.H.: Practical Approach to Electroencephalography. Elsevier Health Sciences, Philadelphia, PA (2012)
26. DeLeeuw, K.E., Mayer, R.E.: A comparison of three measures of cognitive load: Evidence for separable measures of intrinsic, extraneous, and germane load. J. Educ. Psychol. **100**, 223 (2008)

Using Contactless Heart Rate Measurements for Real-Time Assessment of Affective States

Philipp V. Rouast, Marc T.P. Adam, David J. Cornforth, Ewa Lux, and Christof Weinhardt

Abstract Heart rate measurements contain valuable information about a person's affective state. There is a wide range of application domains for heart rate-based measures in information systems. To date, heart rate is typically measured using skin contact methods, where users must wear a measuring device. A non-contact and easy to use mobile approach, allowing heart rate measurements without interfering with the users' natural environment, could prove to be a valuable NeuroIS tool. Hence, our two research objectives are (1) to develop an application for mobile devices that allows for non-contact, real-time heart rate measurement and (2) to evaluate this application in an IS context by benchmarking the results of our approach against established measurements. The proposed algorithm is based on non-contact photoplethysmography and hence takes advantage of slight skin color variations that occurs periodically with the user's pulse.

Keywords Heart rate • Photoplethysmography • Mobile • NeuroIS • Information systems

1 Introduction

Affective states are increasingly gaining attention within information systems (IS) research [1]. They provide valuable insights for the evaluation and evolution of IS related domains such as human-computer interaction (HCI) or decision support systems (DSS). Among a range of different tools in NeuroIS research, heart rate (HR) measurements contribute to a deeper understanding of cognitive and affective processes in IS [1, 2]. For example, HR measurements have been used to evaluate the impact of computerized agents in electronic auctions [3], to detect emotions in financial decision making [4] or as neurophysiological correlates for investigating

P.V. Rouast • E. Lux (✉) • C. Weinhardt
Karlsruhe Institute of Technology, Karlsruhe, Germany
e-mail: philipp.rouast@student.kit.edu; ewa.lux@kit.edu; christof.weinhardt@kit.edu

M.T.P. Adam • D.J. Cornforth
University of Newcastle, Newcastle, Australia
e-mail: marc.adam@newcastle.edu.au; david.cornforth@newcastle.edu.au

F.D. Davis et al. (eds.), *Information Systems and Neuroscience*, Lecture Notes in Information Systems and Organisation 16, DOI 10.1007/978-3-319-41402-7_20

cognitive absorption in enactive training [5]. Hence, there is a great potential for integrating HR measurements as real-time input in various IS domains, e.g., technostress applications [6], e-learning systems [7], and electronic auctions [8].

Recent research investigates increasingly unobtrusive NeuroIS tools for measuring affective states, such as standard mouse devices [9], and aims at designing innovative and useful IT artifacts [10]. While HR data is typically collected using skin contact measurement methods, such as electronic or optical sensors, new algorithms are developed to measure human HR without skin contact by analyzing video recordings [11]. Such techniques fall into the field of non-contact photoplethysmography (PPG) and utilize phenomena such as the periodic variation of humans' skin color to provide reliable methods for measuring HR, without the need to attach sensors to the user's skin. Since a large portion of recent work on non-contact PPG investigates feasibility under lab conditions using previously recorded video, research on mobile real-time applications is scarce.

In this paper, we aim to develop and evaluate an approach for real-time non-contact HR measurements using mobile devices. Our first research objective is the design of an artifact that builds on existing algorithms for non-contact PPG and enables real-time measurements on a mobile device. Our second research objective is the evaluation of the developed artifact in a NeuroIS study by benchmarking it against standard electrocardiogram (ECG) recordings and evaluating limitations and benefits of the developed algorithm for further IS research.

2 Theoretical Background

Recent research shows that remote video recordings of subjects under ambient light contain a rich enough signal to measure vital functions such as HR and respiration rate [12]. Based on the framework of the biological measuring chain [13] we subdivide existing approaches of non-contact PPG for HR measurement into three key steps: (1) signal extraction from a series of images, (2) filtering of the signal to eliminate noise, and (3) estimation of the heart beat frequency (i.e. HR). In the following, we review approaches for non-contact PPG and evaluate their appropriateness for IS research with respect to these three steps. They serve as reference and foundation for the proposed real-time, mobile approach to unobtrusively measure HR in IS settings.

Signal Extraction Two main approaches for HR measurement dominate the existing literature on non-contact PPG: HR measurement based on head movement and HR measurement based on color variation. The first approach utilizes periodic motion of the head caused by the opposite reaction of blood being pumped into the head through the carotid arteries. Hence, it requires the head to remain stationary [14–16]. In a mobile approach, it is likely that the user will move. This is why we focus on the second approach, HR measurement based on color variation, which utilizes the subtle change in the skin color of the human face that occurs periodically with the pulse. Invisible to the human eye, this variation can be used to measure the HR using video analysis algorithms [11, 12, 17–20]. The first step is to select the face as the region of interest (ROI). Existing approaches for ROI detection range from manual classification [12, 18] to computationally expensive feature point detection algorithms [20]. For real-time application, the use of a simple classification algorithm is an appropriate trade-off between complexity and accuracy, permitting an automated ROI detection. Most previous work relies on such algorithms [11, 14, 15, 19, 20].

To eliminate the noise introduced by rigid body movements, the user's face needs to be tracked and the ROI updated. If facial feature points are available, an implementation of an optical flow estimation algorithm can be used [20]. To eliminate non-rigid movements, caused by natural movements in the user's face, areas such as the eyes [20] are usually excluded from the ROI, or the ROI is defined as a rectangle on the user's forehead [18]. The raw signal is then obtained by computing the average value of color channels within the ROI [11, 12, 18, 20]. Whether to use all or rather a subset of color channels from the plethysmographic signal in the subsequent analysis constitutes a trade-off between complexity and information [11, 12, 18].

Filtering The raw signal can contain high frequency noise (e.g., digital artifacts) and low frequency noise (e.g., illumination changes). Digital filters are usually applied to the raw signal to increase the ratio of signal to noise. The main goal is to emphasize the feasible frequencies for human HR. A moving average filter can be applied to eliminate low frequency noise [15]. Since the HR of a healthy human can be assumed to lie within a certain bandwidth of frequencies, a bandpass filter (e.g., a Butterworth filter) is applied to the signal by many authors [15, 16, 18] to remove the unwanted low- and high frequency noise. [20] use the advanced detrending filter proposed by [21] to remove the long-running trend of the raw signal. In the case of a multidimensional signal, a dimensionality reduction technique is applied to the signals to isolate the desired part of the signal. This can be achieved using the Principal Component Analysis (PCA) [14, 18]. As dimensionality reduction techniques produce multiple components, a way to identify the most periodic component has to be specified, e.g., the component with the highest percentage of spectral power accounted by the first harmonic after a Discrete Fourier Transform (DFT) [16].

Frequency Estimation The HR is extracted from the filtered signal using frequency estimation. Most authors use the DFT [11, 12, 16–18] which converts the filtered signal from the time domain to the frequency domain. The HR can then be estimated from the index of the maximum power response. If the individual beat-by-beat intervals are of interest, a peak detection algorithm should be considered, such as in [22].

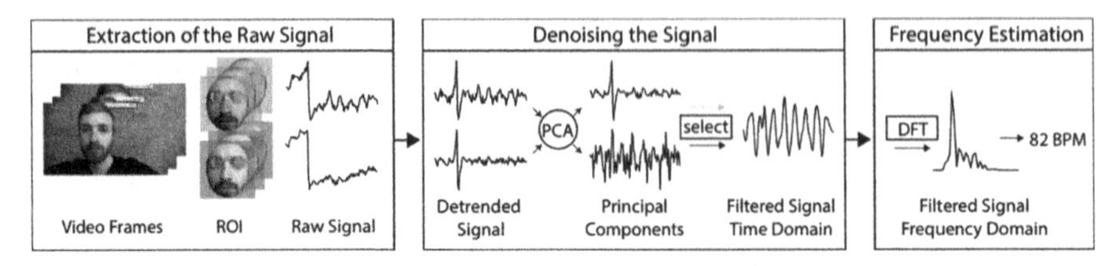

Fig. 1 Overview of our non-contact PPG algorithm approach

3 Approach

According to our first research objective, we developed a non-contact PPG algorithm. The procedure of this algorithm is depicted in Fig. 1. The algorithm is designed to run on mobile devices, hence, we strive for high accuracy, given the circumstances of a typical IS setting. Since we assume that the user interacts with an IS and hence, focuses on a screen, we expect a moderate amount of head movement. As real-time computation on a mobile device puts constraints on complexity and time, trade-offs with the amount of information processed are unavoidable. Through parameterization and scalability, we allow for more accurate results in the future as more powerful mobile devices emerge. Additionally, further features such as HR variability, as proposed by [23], could be integrated into the algorithm.

For the first step, the ROI is defined. We use the popular Viola-Jones object detector to obtain a bounding box for the face and both eyes of the subject [24]. The ROI comprises an ellipse around the center of the bounding box of the face with 80% of its width and 100% of its height, excluding two circles around the centers of the eyes. This way we are able to capture the subject's face with little noise from background pixels and eye blinking. Faces are re-detected at an adjustable time interval[1] and matched to achieve a quasi-tracking ability. From each frame masked to the ROI, we then spatially pool the colors by choosing the mean values of the red and green channels, similar to the approaches of [11, 18]. We decide to use only the red and green channels as a trade-off between complexity and accuracy following [18]. These values constitute the raw signal.

The second step addresses the removal of noise from the raw signal. In preparation for the PCA, we use the detrending filter proposed by [21] to remove any general trend in the data and isolate rapid leaps in the signal caused by face re-detection. The PCA is then applied to the two-dimensional signal, yielding two principal components. One principal component contains HR information, the other noise due to, e.g., lighting changes or rigid body movements. To identify the principal component that represents the HR signal, we adopt the approach of [16] and choose the one with the more distinct periodicity. Since the signal still contains interfering frequencies, we apply a Butterworth bandpass filter [12, 14–16] with the

[1]We use a default interval of one second for tracking.

cut-off frequencies 45 BPM and 240 BPM [16]. In addition, we use a Gaussian smoothing filter to reduce the noise and yield a filtered, one-dimensional signal.

The third step is to estimate the frequency of the heartbeat. For this purpose, we again transform the signal to the frequency domain using a DFT and find the index with the highest spectral power. Using this index, we are able to estimate the HR.

4 Further Research Agenda

Information about users' affective states provide valuable insights for IS research [3, 5, 25]. As HR is an important indicator of emotional arousal, HR measurements are a valuable NeuroIS tool [1, 2]. However, HR measurements are often performed using ECG or contact PPG, where the user has to wear sensors that require skin contact (i.e., ECG electrodes or finger clip/wristband with optical sensor). To make HR measurements less obtrusive, we addressed the first of our two research objectives in this paper. A non-contact mobile approach for real-time HR measurement is discussed and an algorithm based on recent work on non-contact PPG is developed. Our approach aims at enabling ubiquitous and universal accessibility of HR measurements and could be utilized in the future as real-time system input for both, the evaluation and adaptation of IS. Moreover, recording the data on the user's own smartphone can be considered useful to address privacy protection concerns related to the individual's physiological data. In particular, [6, 26] argued that physiological data of a person should be owned by the respective person. Processing physiological data on their own smartphone enables the user to own this sensitive data and to have direct control over who has access to it.

With respect to the second research objective, we will evaluate this approach by conducting a NeuroIS study similar to the study of [8]. The aim of the planned evaluation study is to benchmark the non-contact real-time measurements of the proposed mobile approach against established ECG HR measurements and evaluate its usefulness for IS research. A study such as [8] would be suitable for the evaluation of the proposed algorithm as it contains elements such as user interaction, social competition, and time pressure, which are likely to be relevant for future studies that include affective information. In this study, the users' affective state is assessed using ECG measurements, which require three skin electrodes on the user's chest. To transfer this lab experiment to the field or to provide users with real-time feedback about their affective states, [8] request a mobile, non-contact solution that extracts information from the acquired data in real-time. Next to the scenario investigated by [8], the proposed NeuroIS tool could be used to provide traders and investors with a direct feedback on their emotional state [27–29] or to support upcoming remedies for technostress [25].

In the evaluation study, we will investigate the accuracy of our proposed PPG-based approach in different scenarios: Resting, low time pressure auction, and high time pressure auction. The study will contain the same elements as the study by [8]. It will be conducted in a lab equipped with a computer terminal for each client. Prior to the auctions, an initial rest period of five minutes will be conducted, during which the participants are asked to relax, so that their physiology can adjust to a resting level. We expect that the PPG-based video image analysis for HR measurements yields the highest accuracy in the resting period section. During the remainder of the experiment, the users' HRs are measured in electronic auctions with low and high levels of time pressure in order to investigate how the accuracy of the non-contact PPG-based approach varies in different settings. As a result of the evaluation study, we will benchmark the measurements based on the proposed application against the measurements by the established ECG technology. We plan to compare the two techniques with respect to subjective as well as objective criteria. Subjective evaluation criteria will depend on the users' perceptions and determine to what extent the users felt distracted by the measurements, objective evaluation criteria will be the correctness of the HR detection. Finally, we will evaluate users' general level of technology acceptance of the proposed measurement approach.

References

1. Riedl, R., Davis, F.D., Hevner, A.R.: Towards a NeuroIS research methodology: Intensifying the discussion on methods, tools, and measurement. J. Assoc. Inf. Syst. **15**, i–xxxv (2014)
2. Dimoka, A., Banker, R.D., Benbasat, I., Davis, F.D., Dennis, A.R., Gefen, D., Gupta, A., Ischebeck, A., Kenning, P., Pavlou, P.A., Müller-Putz, G., Riedl, R., vom Brocke, J., Weber, B.: On the use of neurophysiological tools in IS research: Developing a research agenda for NeuroIS. MIS Q. **36**, 679–702 (2012)
3. Teubner, T., Adam, M.T.P., Riordan, R.: The impact of computerized agents on immediate emotions, overall arousal and bidding behavior in electronic auctions. J. Assoc. Inf. Syst. **16**, 838–879 (2015)
4. Hariharan, A., Adam, M.T.P.: Blended emotion detection for decision support. IEEE Trans. Hum.-Mach. Syst. **45**, 510–517 (2015)
5. Léger, P.-M., Davis, F.D., Cronan, T.P., Perret, J.: Neurophysiological correlates of cognitive absorption in an enactive training context. Comput. Hum. Behav. **34**, 273–283 (2014)
6. Adam, M.T.P., Gimpel, H., Maedche, A., Riedl, R.: Design blueprint for stress-sensitive adaptive enterprise systems. Bus. Inf. Syst. Eng. (in press)
7. Shen, L., Wang, M., Shen, R.: Affective e-learning: Using "emotional" data to improve learning in pervasive learning environment. Educ. Technol. Soc. **12**, 176–189 (2009)
8. Adam, M.T.P., Krämer, J., Müller, M.B.: Auction fever! How time pressure and social competition affect bidders' arousal and bids in retail auctions. J. Retail. **91**, 468–485 (2015)
9. Schaaff, K., Degen, R., Adler, N., Adam, M.T.P.: Measuring affect using a standard mouse device. Biomed. Eng. (NY) **57**, 761–764 (2012)
10. Vom Brocke, J., Riedl, R., Léger, P.-M.: Application strategies for neuroscience in information systems design science research. J. Comput. Inf. Syst. **53**, 1–13 (2013)
11. Poh, M.-Z., McDuff, D.J., Picard, R.W.: Non-contact, automated cardiac pulse measurements using video imaging and blind source separation. Opt. Express. **18**, 10762–10774 (2010)
12. Verkruysse, W., Svaasand, L.O., Nelson, J.S.: Remote plethysmographic imaging using ambient light. Opt. Express. **16**, 21434–21445 (2008)
13. Hoffmann, K.-P.: Biosignale Erfassen und Verarbeiten. In: Kramme, R. (ed.) Medizintechnik, pp. 667–688. Springer, Berlin (2011)

14. Balakrishnan, G., Durand, F., Guttag, J.: Detecting pulse from head motions in video. In: Proceedings of the 2013 I.E. Computer Society Conference on Computer Vision and Pattern Recognition, pp. 3430–3437 (2013)
15. Irani, R., Nasrollahi, K., Moeslund, T.B.: Improved pulse detection from head motions using DCT. In: Proceedings of the 9th International Conference on Computer Vision Theory and Applications, pp. 118–124 (2014)
16. Shan, L., Yu, M.: Video-based heart rate measurement using head motion tracking and ICA. In: Proceedings of the 2013 6th International Congress on Image and Signal Processing, pp. 160–164 (2013)
17. Wu, H.-Y., Rubinstein, M., Shih, E., Guttag, J.V., Durand, F., Freeman, W.T.: Eulerian video magnification for revealing subtle changes in the world. ACM Trans. Graph. **31**, 1–8 (2012)
18. Lewandowska, M., Ruminski, J., Kocejko, T.: Measuring pulse rate with a webcam: A non-contact method for evaluating cardiac activity. In: Proceedings of the 2011 Federated Conference on Computer Science and Information Systems (FedCSIS), pp. 405–410 (2011)
19. Kwon, S., Kim, H., Park, K.S.: Validation of heart rate extraction using video imaging on a built-in camera system of a smartphone. In: Proceedings of the 2012 I.E. Annual International Conference of the Engineering in Medicine and Biology Society, pp. 2174–2177 (2012)
20. Li, X., Chen, J., Zhao, G., Pietikäinen, M.: Remote heart rate measurement from face videos under realistic situations. In: Proceedings of the 2014 I.E. Computer Society Conference on Computer Vision and Pattern Recognition, pp. 4264–4271 (2014)
21. Tarvainen, M.P., Ranta-Aho, P.O., Karjalainen, P.A.: An advanced detrending method with application to HRV analysis. IEEE Trans. Biomed. Eng. **49**, 172–175 (2002)
22. Wu, H.-Y.: Eulerian Video Processing and Medical Applications. Master's Thesis, Massachusetts Institute of Technology (2012)
23. Müller, M.B., Adam, M.T.P., Cornforth, D.J., Chiong, R., Krämer, J., Weinhardt, C.: Selecting physiological features for predicting bidding behavior in electronic auctions. In: Proceedings of the Forty-Ninth Annual Hawaii International Conference on System Sciences (HICSS), pp. 396–405 (2016)
24. Viola, P., Jones, M.: Rapid object detection using a boosted cascade of simple features. In: Proceedings of the 2001 I.E. Computer Society Conference on Computer Vision and Pattern Recognition, pp. 511–518 (2001)
25. Riedl, R.: On the biology of technostress: Literature review and research agenda. ACM SIGMIS Database **44**, 18–55 (2013)
26. Fairclough, S.: Physiological data must remain confidential. Nature **505**, 263 (2014)
27. Astor, P.J., Adam, M.T.P., Jerčić, P., Schaaff, K., Weinhardt, C.: Integrating biosignals into information systems: A NeuroIS tool for improving emotion regulation. J. Manag. Inf. Syst. **30**, 247–278 (2013)
28. Lux, E., Hawlitschek, F., Adam, M.T.P., Pfeiffer, J.: Using live biofeedback for decision support: Investigating influences of emotion regulation in financial decision making. In: ECIS 2015 Research-in-Progress Papers, pp. 1–12 (2015)
29. Astor, P.J., Adam, M.T.P., Jähnig, C., Seifert, S.: Measuring regret: Emotional aspects of auction design. In: ECIS 2011 Proceedings, pp. 1129–1140 (2011)

Lifelogging as a Viable Data Source for NeuroIS Researchers: A Review of Neurophysiological Data Types Collected in the Lifelogging Literature

Thomas Fischer and René Riedl

Abstract Based on this review, we argue for the consideration of lifelogging as an additional data source in NeuroIS research. Lifelogging itself is a concept which describes a behavior in which individuals, based on the use of computer technology, track (parts of) their lives, including the quantification of their well-being (e.g., continuous recording of an individual's heart rate via a digital wrist watch). This relatively new form of behavior generates a viable data source for future NeuroIS studies, predominantly for those conducted in field settings. By analyzing how frequently the major types of neurophysiological data have thus far been collected in lifelogging publications, we reveal how much attention different types of neurophysiological data have received in the context of longitudinal field studies. In essence, lifelogging data constitute a viable data base for NeuroIS researchers, one that is readily available and is predicted to grow in the future because an increasing number of people worldwide are tracking their daily lives to a growing extent.

Keywords Field studies • Lifelogging • NeuroIS • Self-tracking

1 Introduction

It has previously been highlighted that there is a lack of NeuroIS studies in the field [1]. While investigations in field settings may imply increased research complexity for several reasons (e.g., intrusion of research into private areas), there is a clear need for the investigation of focal constructs, such as stress, in the daily life of users. In this paper, we highlight the value of data collected by individuals

T. Fischer (✉)
University of Applied Sciences Upper Austria, Steyr, Austria
e-mail: thomas.fischer@fh-steyr.at

R. Riedl
University of Appl. Sc. Upper Austria, Steyr, Oberösterreich, Austria

University of Linz, Linz, Austria
e-mail: rene.riedl@fh-steyr.at

F.D. Davis et al. (eds.), *Information Systems and Neuroscience*, Lecture Notes in Information Systems and Organisation 16, DOI 10.1007/978-3-319-41402-7_21

themselves, a behavior which is typically referred to as lifelogging. Thus, lifelogging data constitute a new and readily available data source for NeuroIS researchers.

Lifelogging has the goal of enabling an individual to collect the totality of her/his experiences through the digitization of all cognitive inputs and/or neurophysiological activation [2]. Contemporary lifelogging research has its roots in an early vision of Vannevar Bush [3] who proposed a device which would allow for the capture and storage of all information that an individual would amass over a lifetime. Ongoing developments related to sensor and storage technologies make the realization of such a vision possible (e.g., [2, 4–7]).

In the next section, we briefly discuss the concept of lifelogging and highlight previous calls for the use of neurophysiological data in lifelogging research. Afterwards, we review the literature to reveal how much attention different types of neurophysiological data have received in the context of longitudinal lifelogging field studies.

2 Lifelogging

As defined by Dodge and Kitchin [2], a "life-log" is "(. . .) a form of pervasive computing consisting of a unified digital record of the *totality* of an individual's experiences, captured multimodally through digital sensors and stored permanently as a personal multimedia archive" (p. 431, italics in original). Hence, lifelogging tools are intended to penetrate an individual's daily life, collecting data on a longitudinal basis. Though not in the propagated "total" degree, Swan [8] highlighted that such self-tracking behaviors are already an integral part of our life, as many individuals regularly measure something about themselves, in part due to an intrinsic motivation to figure out their inner workings (e.g., related to dietary choices). From a NeuroIS standpoint, health and fitness-related applications, among others, are highly interesting because the type of data collected in these application domains is related to important outcome variables in IS research, such as stress [9, 10] or technostress [11, 12]. Health and fitness-related applications are currently a dominant domain in lifelogging research [13].

Lifelogging could be of interest to an even wider public, and, importantly, often does not involve a significant change in an individual's daily behavior. In a 2013 survey in the United States, 69% of the participants already reported that they were regularly keeping track of at least one health indicator (e.g., diet or weight) though only 21 % used technology to support this endeavor [14]. In a 2014 survey on the same topic, the share of self-trackers in the US public was even higher, with an astounding 91 % [15]. Powerful wearable technologies are supporting the increasingly effortless collection of lifelogging data, with IDC estimating that sales in this category will surge from almost 20 million units in 2014 to more than 126 million in 2019, with wrist wear being the dominant type of device in this context (around 80–90 % of sales) [16]. In addition, researchers who have been interviewed as part of the more recent survey [15] stated that they are most interested in data specifically related to the well-being of individuals such as vital signs, as well as data on physical activity, mood, and stress.

Importantly, there have been several calls for lifelogging research to more actively focus on the capture of the internal mechanisms of individuals. For example, Nack [17] highlighted that lifelogging has been centered too much around the idea of creating an additional memory, capturing the external context, while widely neglecting the internal context (e.g., affective states and physiological processes). Hence, it has been argued that also unconsciously processed data involved in the physiological functioning of individuals constitute a valuable addition to lifelogging applications (e.g., [2, 18, 19]).

Due to the enormous potential that lifelogging applications hold for NeuroIS research, particularly for investigations conducted in the field, it is important to assess the role of neurophysiological data in lifelogging publications.

3 Neurophysiological Data in Lifelogging Publications

To assess the prevalence of neurophysiological data in lifelogging research, we conducted a literature review based on the following criteria. We used the term "lifelog" in Google Scholar (02/22/2016: 5050 hits) and selected journal publications as well as publications in scientific periodicals which had accumulated more than five citations thus far.[1] These criteria led to a set of 54 publications, but 16 papers were excluded as they did not offer sufficient information on the types of data involved. Two additional articles were excluded (one only involved paper-based logging of information and the other one involved the collection of personal health information, though only for the use of medical personnel (i.e., a form of monitoring, hence not an application of individual lifelogging).[2]

As lifelogging aims to be all-encompassing, the types of data that are captured can be almost limitless (e.g., [4, 13, 21]). Therefore, we employed an existing categorization scheme introduced by Jacquemard et al. [22] to divide them into four main groups: data on the individual, data on the environment, device-specific data, and third-party data (see Table 1).

The first category is "inward facing", which we describe with *data on the individual.* By far the most popular source of data on the individual in the reviewed publications was motion, measured via accelerometers which can be indicative of physical activity levels, sedentary behaviors, and can even be used to calculate energy expenditures. Additionally, we highlighted studies which utilized the SenseCam, a camera device that has been widely prevalent in lifelogging research.

[1]A similar criterion has previously been applied in a literature review by Riedl [20].

[2]It has to be noted that we do not claim that the 36 publications which we identified constitute an exhaustive list of available publications.

Table 1 Prevalence of types of data employed in lifelogging publications

Studies	Citations[a]	Individual	Environment	Device-specific	Third-Party
Gemmell et al. (2006) [23]	517	X	X	X	X
Choudhury et al. (2008) [24]	400	X	X		
Anderson et al. (2007) [25]	167			X	
Berry et al. (2007) [26]	165	(*)	X		
Gyorbiro et al. (2009) [27]	154	X	X	X	
Whittaker et al. (2010) [28]	80		X		
Blum et al. (2006) [29]	78	X	X	X	
Hodges et al. (2011) [30]	68	(*)	X		
Lee et al. (2008) [31]	67	(*)	X		
Ogata et al. (2011) [32]	66	X	X		
Doherty et al. (2011) [33]	63	(*)	X		
Kelly et al. (2011) [34]	62	(*)	X		
Cho et al. (2007) [35]	59		X	X	X
Whittaker et al. (2008) [36]	50		X		
Jacques et al. (2011) [37]	49	(*)	X		
Doherty et al. (2011) [38]	45	(*)	X		
Vemuri and Bender (2004) [39]	43		X	X	X
Berry et al. (2009) [40]	43	(*)	X		
Lee et al. (2011) [41]	42	X			
Doherty et al. (2012) [42]	40	(*)	X		
Hwang and Cho (2009) [43]	38		X	X	X
Whittaker et al. (2012) [44]	34	X	X		
Rawassizadeh et al. (2013) [45]	33	X	X	X	
Gurrin et al. (2013) [46]	32	X	X	X	
Abe et al. (2009) [47]	31	X	X		
Byrne et al. (2010) [48]	31	(*)	X		
Browne et al. (2011) [49]	29	(*)	X		
Doherty et al. (2010) [50]	28	(*)	X		X
Pauly-Takacs et al. (2011) [51]	26	(*)	X		
Kikhia et al. (2010) [52]	25		X		
Brindley et al. (2011) [53]	23	X	X		
Ryoo and Bae (2007) [54]	20	X	X		
Ivonin et al. (2013) [19]	20	X	X		

(continued)

Table 1 (continued)

Studies	Citations[a]	Individual	Environment	Device-specific	Third-Party
Kalnikaite et al. (2011) [55]	18	X	X		
Ogata et al. (2008) [56]	15		X		
Wang et al. (2012) [57]	12	(*)	X		

[a]Assessed via Google Scholar on 02/22/2016
(*) = used the SenseCam, but did not use its integrated accelerometer for data collection

The SenseCam is a wearable camera with a number of integrated sensors that takes images from an individual's point of view every few seconds or even at shorter intervals if the sensors detect a significant change in the environmental setting or changes in an individual's posture (e.g., [30, 58]).

The second category of "outward facing" data includes *data on the environment* which can involve all types of contextual information. Visual lifelogs are by far the most prominent data source in this category, mostly applying the SenseCam for data collection or similar wearable camera devices.

The third category of *device-specific data* includes all kinds of data about information systems and their functioning (e.g., received phone calls, battery status, or used applications).

Finally, in the fourth category of *third-party data* we include data on the individual that has not actually been collected by the individual herself or himself and can be requested from others (e.g., medical data). In the selected studies, only weather data was collected.

From a NeuroIS standpoint, it is important to note that only three of the reviewed studies (i.e., [19, 53, 54]) included data which is amongst the "common" types of data proposed for NeuroIS (e.g., [59, 60]). Ryoo and Bae [54] presented the concept for a wearable gadget that could also collect cardiovascular activity via ECG and infer stress levels from the gathered signals. Ivonin et al. [19] presented a method to recognize what they labelled "unconscious emotions" from heart signals as an extension for future lifelogging applications. Finally, Brindley et al. [53] used SenseCam to collect visual data, and, in addition, gathered affective data using an ECG monitor on a patient with acquired brain injury. They tested whether data collected during social situations in the field would elicit similar responses in the patient when showing him SenseCam pictures of the same events again in the laboratory.

A potential explanation for the seemingly low popularity of other neurophysiological data in these lifelogging studies is the requirement for lifelogging applications to be effortless [7] (as they should essentially be used throughout the day and ideally capture a whole lifetime), and they therefore need to be particularly unobtrusive. Riedl et al. [61] emphasized that the criterion of intrusiveness is amongst the most important factors in NeuroIS measurement, highlighting that, ideally, neurophysiological tools should allow for a high degree of movement freedom, high degree of natural position, and a low degree of invasiveness. It follows that a number of neurophysiological tools (e.g., fMRI) are not applicable in a lifelogging context.

4 Future Directions

Pantelopoulos and Bourbakis [62] and Appelboom et al. [63] listed a number of biological sensors that can be used to gather individual-level data in field settings which not only included cardiovascular activity, but also blood pressure, body and/or skin temperature, respiration rate, oxygen saturation, electro-dermal activity, blood glucose values, muscular activity via electromyogram, or brain activity via electroencephalogram (EEG). Hence, through back-search in our selected publications (see Table 1) we identified a number of inspiring applications which show that lifelogging might actually more strongly embrace neurophysiological signals, which would make this research domain even more interesting from a NeuroIS perspective.

In an early study, Healey and Picard [64] used the startle response, inferred from electro-dermal activity, to create the "StartleCam", a wearable camera which would mainly collect moments that are likely valuable to the individual and hence elicit an affective response. Hori and Aizawa [65] even included a brain-wave analyzer (EEG) in their lifelogging application which may be used to collect alpha waves that would then be informative of an individual's attention levels, which could be used, for example, to more effectively retrieve interesting scenes from a continuous video-log. The array of sensors that could be put on an individual was even further evolved by Matthews et al. [66] who developed a physiological sensor suite (PSS) that could include several modules of physiological sensors including ECG, EEG, electromyogram (muscular activity), and electrooculogram (eye movements).

Aside from being worn by the individual, relevant sensors could also be integrated into the environment, creating ambient applications. For example, Healey and Picard [67] created a set-up which would allow for the collection of physiological data informative of individual stress levels (i.e., cardiovascular activity via ECG, muscular activity via EMG, skin conductance and respiration) inside a car and tested it in an actual field course. Finally, in an organizational setting, McDuff et al. [68] proposed a set-up which would allow employees to track their daily work activities including a WebCam, Microsoft Kinect to capture body movements, microphone, GPS sensor, capture of file activity on the local workstation, and a wrist-worn EDA sensor. Also, Schaaff et al. [69] presented a prototype of a computer mouse equipped with sensors which may detect pressure from individuals interacting with the mouse that could then be used to infer emotional states (e.g., levels of frustration).

Despite the fact that these studies create a more versatile picture of the application of neurophysiological data in lifelogging, they also highlight that a main challenge to the application of such types of data in lifelogging contexts is the number of confounding variables that have to be considered (e.g., [53, 66, 67]), such as noise (e.g., muscle contractions) created from fast movements. This fact has also been pointed out by Appelboom et al. [63] in the context of tele-monitoring, who proposed that the ideal application would have to essentially be tailored to each individual interested in self-tracking, hence requiring a high level of self-interest in the tracking of physiological data by individuals.

In conclusion, while there are applications that show that neurophysiological data can be collected in the field in a longitudinal fashion, the creation of a life-long archive by individuals themselves requires an approach that is mostly driven by the individual herself/himself, though should be supported by researchers interested in the gathered data. Thus, by more actively involving themselves into the development of data collection approaches in the lifelogging domain, NeuroIS researchers could more effectively present their topics (application areas) to individuals interested in self-tracking, and, in addition, create trust in individuals, which has been shown to be an essential critical prerequisite of lifelogging data being shared with interested researchers (e.g., [15]).

References

1. Fischer, T., Riedl, R.: Neurois in situ: On the Need for NeuroIS Research in the Field to Study Organizational Phenomena. In: Liang, T.-P., Yen, N.-S. (eds.) Workshop on Information and Neural Decision Sciences, pp. 20–21 (2014)
2. Dodge, M., Kitchin, R.: 'Outlines of a world coming into existence': Pervasive computing and the ethics of forgetting. Environ. Plann. Plann. Des. **34**, 431–445 (2007)
3. Bush, V.: As we may think. The Atlantic Monthly, pp. 112–124 (1945)
4. Gurrin, C., Smeaton, A.F., Doherty, A.R.: LifeLogging: Personal big data. Found. Trends Inf. Retr. **8**, 1–125 (2014)
5. Bell, G.: A personal digital store. Comm. ACM **44**, 86–91 (2001)
6. O'Hara, K., Morris, R., Shadbolt, N., Hitch, G.J., Hall, W., Beagrie, N.: Memories for life: A review of the science and technology. J. R. Soc. Interface **3**, 351–365 (2006)
7. Sellen, A.J., Whittaker, S.: Beyond total capture. Comm. ACM **53**, 70 (2010)
8. Swan, M.: The quantified self: Fundamental disruption in big data science and biological discovery. Big Data **1**, 85–99 (2013)
9. Weiss, M.: Effects of work stress and social support on information systems managers. MIS Q. **7**, 29 (1983)
10. Sethi, V., King, R.C., Quick, J.C.: What causes stress in information system professionals? Comm. ACM **47**, 99–102 (2004)
11. Riedl, R., Kindermann, H., Auinger, A., Javor, A.: Technostress from a neurobiological perspective—System breakdown increases the stress hormone cortisol in computer users. Bus. Inf. Syst. Eng. **4**, 61–69 (2012)
12. Riedl, R., Kindermann, H., Auinger, A., Javor, A.: Computer breakdown as a stress factor during task completion under time pressure: Identifying gender differences based on skin conductance. Adv. Hum.-Comput. Interact. **2013**, 1–8 (2013)
13. Wang, P., Smeaton, A.F.: Using visual lifelogs to automatically characterize everyday activities. Inf. Sci. **230**, 147–161 (2013)
14. Fox, S., Duggan, M.: Tracking for Health. Pew Research Center's Internet & American Life Project. Pew Research Center (2013)

15. California Institute for Telecommunications and Information Technology: Personal Data for the Public Good. New Opportunities to Enrich Understanding of Individual and Population Health (2014)
16. Worldwide Wearables Market Forecast to Reach 45.7 Million Units Shipped in 2015 and 126.1 Million Units in 2019, According to IDC (2015)
17. Nack, F.: You must remember this. IEEE Multimed. **12**, 4–7 (2005)
18. Bell, G., Gemmell, J.: A digital life. Sci. Am. **296**, 58–65 (2007)
19. Ivonin, L., Chang, H.-M., Chen, W., Rauterberg, M.: Unconscious emotions: Quantifying and logging something we are not aware of. Pers. Ubiquit. Comput. **17**, 663–673 (2013)
20. Riedl, R.: On the biology of technostress: Literature review and research agenda. Database Adv. Inform. Syst. **44**, 18–55 (2013)
21. O'Hara, K., Tuffield, M.M., Shadbolt, N.: Lifelogging: Privacy and empowerment with memories for life. Ident. Inform. Soc. **1**, 155–172 (2008)
22. Jacquemard, T., Novitzky, P., O'Brolcháin, F., Smeaton, A.F., Gordijn, B.: Challenges and opportunities of lifelog technologies: A literature review and critical analysis. Sci. Eng. Ethics **20**, 379–409 (2014)
23. Gemmell, J., Bell, G., Lueder, R.: MyLifeBits: A personal database for everything. Comm. ACM **49**, 88–95 (2006)
24. Choudhury, T., Borriello, G., Consolvo, S., Haehnel, D., Harrison, B., Hemingway, B., Hightower, J., Klasnja, P., Koscher, K., LaMarca, A., et al.: The mobile sensing platform: An embedded activity recognition system. IEEE Pervasive Comput. **7**, 32–41 (2008)
25. Anderson, I., Maitland, J., Sherwood, S., Barkhuus, L., Chalmers, M., Hall, M., Brown, B., Muller, H.: Shakra: Tracking and sharing daily activity levels with unaugmented mobile phones. Mobile Network Appl. **12**, 185–199 (2007)
26. Berry, E., Kapur, N., Williams, L., Hodges, S.E., Watson, P., Smyth, G., Srinivasan, J., Smith, R., Wilson, B., Wood, K.: The use of a wearable camera, SenseCam, as a pictorial diary to improve autobiographical memory in a patient with limbic encephalitis: A preliminary report. Neuropsychol. Rehabil. **17**, 582–601 (2007)
27. Gyorbiro, N., Fabian, A., Hományi, G.: An activity recognition system for mobile phones. Mobile Network Appl. **14**, 82–91 (2009)
28. Whittaker, S., Bergman, O., Clough, P.: Easy on that trigger dad: A study of long term family photo retrieval. Pers. Ubiquit. Comput. **14**, 31–43 (2010)
29. Blum, M., Pentland, A., Troster, G.: InSense: Interest-based life logging. IEEE Multimed. **13**, 40–48 (2006)
30. Hodges, S., Berry, E., Wood, K.: SenseCam: A wearable camera that stimulates and rehabilitates autobiographical memory. Memory **19**, 685–696 (2011)
31. Lee, H., Smeaton, A.F., O'Connor, N.E., Jones, G., Blighe, M., Byrne, D., Doherty, A., Gurrin, C.: Constructing a SenseCam visual diary as a media process. Multimed. Syst. **14**, 341–349 (2008)
32. Ogata, H., Li, M., Hou, B., Uosaki, N., El-Bishouty, M.M., Yano, Y.: SCROLL: Supporting to share and reuse ubiquitous learning log in the context of language learning. Res. Pract. Technol. Enhanc. Learn. **6**, 69–82 (2011)
33. Doherty, A.R., Caprani, N., Conaire, C.Ó., Kalnikaite, V., Gurrin, C., Smeaton, A.F., O'Connor, N.E.: Passively recognising human activities through lifelogging. Comput. Hum. Behav. **27**, 1948–1958 (2011)
34. Kelly, P., Doherty, A.R., Berry, E., Hodges, S.E., Batterham, A.M., Foster, C.: Can we use digital life-log images to investigate active and sedentary travel behaviour? Results from a pilot study. Int. J. Behav. Nutr. Phys. Act. **8**, 44 (2011)
35. Cho, S.-B., Kim, K.-J., Hwang, K.S., Song, I.-J.: AniDiary: Daily cartoon-style diary exploits Bayesian networks. IEEE Pervasive Comput. **6**, 66–75 (2007)
36. Whittaker, S., Tucker, S., Swampillai, K., Laban, R.: Design and evaluation of systems to support interaction capture and retrieval. Pers. Ubiquit. Comput. **12**, 197–221 (2008)

37. Jacques, P.L.S., Conway, M.A., Lowder, M.W., Cabeza, R.: Watching my mind unfold versus yours: An fMRI study using a novel camera technology to examine neural differences in self-projection of self versus other perspectives. J. Cognit. Neurosci. **23**, 1275–1284 (2011)
38. Doherty, A.R., Moulin, C.J.A., Smeaton, A.F.: Automatically assisting human memory: A SenseCam browser. Memory **19**, 785–795 (2011)
39. Vemuri, S., Bender, W.: Next-generation personal memory aids. BT Technol. J. **22**, 125–138 (2004)
40. Berry, E., Hampshire, A., Rowe, J., Hodges, S., Kapur, N., Watson, P., Browne, G., Smyth, G., Wood, K., Owen, A.M.: The neural basis of effective memory therapy in a patient with limbic encephalitis. J. Neurol. Neurosurg. Psych. **80**, 1202–1205 (2009)
41. Lee, M.-W., Khan, A.M., Kim, T.-S.: A single tri-axial accelerometer-based real-time personal life log system capable of human activity recognition and exercise information generation. Pers. Ubiquit. Comput. **15**, 887–898 (2011)
42. Doherty, A.R., Pauly-Takacs, K., Caprani, N., Gurrin, C., Moulin, C.J.A., O'Connor, N.E., Smeaton, A.F.: Experiences of aiding autobiographical memory using the SenseCam. Hum. Comput. Interact. **27**, 151–174 (2012)
43. Hwang, K.-S., Cho, S.-B.: Landmark detection from mobile life log using a modular Bayesian network model. Expert Syst. Appl. **36**, 12065–12076 (2009)
44. Whittaker, S., Kalnikaite, V., Petrelli, D., Sellen, A.J., Villar, N., Bergman, O., Clough, P., Brockmeier, J.: Socio-technical lifelogging: Deriving design principles for a future proof digital past. Hum. Comput. Interact. **27**, 37–62 (2012)
45. Rawassizadeh, R., Tomitsch, M., Wac, K., Tjoa, A.M.: UbiqLog: A generic mobile phone-based life-log framework. Pers. Ubiquit. Comput. **17**, 621–637 (2013)
46. Gurrin, C., Qiu, Z., Hughes, M., Caprani, N., Doherty, A.R., Hodges, S.E., Smeaton, A.F.: The smartphone as a platform for wearable cameras in health research. Am. J. Prev. Med. **44**, 308–313 (2013)
47. Abe, M., Morinishi, Y., Maeda, A., Aoki, M., Inagaki, H.: A life log collector integrated with a remote-controller for enabling user centric services. IEEE Trans. Consum. Electron **55**, 295–302 (2009)
48. Byrne, D., Doherty, A.R., Snoek, C.G.M., Jones, G.J.F., Smeaton, A.F.: Everyday concept detection in visual lifelogs: Validation, relationships and trends. Multimed. Tool. Appl. **49**, 119–144 (2010)
49. Browne, G., Berry, E., Kapur, N., Hodges, S., Smyth, G., Watson, P., Wood, K.: SenseCam improves memory for recent events and quality of life in a patient with memory retrieval difficulties. Memory **19**, 713–722 (2011)
50. Doherty, A.R., Smeaton, A.F.: Automatically augmenting lifelog events using pervasively generated content from millions of people. Sensors **10**, 1423–1446 (2010)
51. Pauly-Takacs, K., Moulin, C.J.A., Estlin, E.J.: SenseCam as a rehabilitation tool in a child with anterograde amnesia. Memory **19**, 705–712 (2011)
52. Kikhia, B., Hallberg, J., Bengtsson, J.E., Savenstedt, S., Synnes, K.: Building digital life stories for memory support. Int. J. Comput. Healthc. **1**, 161–176 (2010)
53. Brindley, R., Bateman, A., Gracey, F.: Exploration of use of SenseCam to support autobiographical memory retrieval within a cognitive-behavioural therapeutic intervention following acquired brain injury. Memory **19**, 745–757 (2011)
54. Ryoo, D.-w., Bae, C.: Design of the wearable gadgets for life-log services based on UTC. IEEE Trans. Consum. Electron **53**, 1477–1482 (2007)
55. Kalnikaite, V., Whittaker, S.: A saunter down memory lane: Digital reflection on personal mementos. Int. J. Hum.-Comput. Stud. **69**, 298–310 (2011)
56. Ogata, H., Misumi, T., Matsuka, T., El-Bishouty, M.M., Yano, Y.: A framework for capturing, sharing and comparing learning experiences in a ubiquitous learning environment. Res. Pract. Technol. Enhanc. Learn. **03**, 297–312 (2008)
57. Wang, P., Smeaton, A.F.: Semantics-based selection of everyday concepts in visual lifelogging. Int. J. Multimedia. Inform. Retrieval **1**, 87–101 (2012)

58. Hodges, S.E., Williams, L., Berry, E., Izadi, S., Srinivasan, J., Butler, A., Smyth, G., Kapur, N., Wood, K.: SenseCam: A retrospective memory aid. In: Hutchison, D., Kanade, T., Kittler, J., Kleinberg, J.M., Mattern, F., Mitchell, J.C., Naor, M., Nierstrasz, O., Pandu Rangan, C., Steffen, B., et al. (eds.) UbiComp 2006: Ubiquitous Computing, 4206, pp. 177–193. Springer, Berlin (2006)
59. Dimoka, A., Banker, R.D., Benbasat, I., Davis, F.D., Dennis, A.R., Gefen, D., Gupta, A., Ischebeck, A., Kenning, P.H., Pavlou, P.A., et al.: On the use of neurophysiological tools in is research: Developing a research agenda for NeuroIS. MIS Q. **36**, 679–702 (2012)
60. Riedl, R., Banker, R.D., Benbasat, I., Davis, F.D., Dennis, A.R., Dimoka, A., Gefen, D., Gupta, A., Ischebeck, A., Kenning, P., et al.: On the foundations of NeuroIS: Reflections on the Gmunden Retreat 2009. Comm. Assoc. Inform. Syst. **27**, 243–264 (2010)
61. Riedl, R., Davis, F.D., Hevner, A.R.: Towards a NeuroIS research methodology: Intensifying the discussion on methods, tools, and measurement. J. Assoc. Inform. Syst. **15**, i–xxxv (2014)
62. Pantelopoulos, A., Bourbakis, N.G.: A survey on wearable sensor-based systems for health monitoring and prognosis. IEEE Trans. Syst. Man Cybern. C Appl. Rev. **40**, 1–12 (2010)
63. Appelboom, G., Camacho, E., Abraham, M.E., Bruce, S.S., Dumont, E.L., Zacharia, B.E., D'Amico, R., Slomian, J., Reginster, J.Y., Bruyère, O., et al.: Smart wearable body sensors for patient self-assessment and monitoring. Arch Public Health **72**, 28 (2014)
64. Healey, J., Picard, R.W.: StartleCam: A cybernetic wearable camera. In: Proceedings of the Second International Symposium on Wearable Computers, pp. 42–49 (1998)
65. Hori, T., Aizawa, K.: Context-based video retrieval system for the life-log applications. In: Sebe, N., Lew, M.S., Djeraba, C. (eds.) Proceedings of the 5th ACM SIGMM International Workshop on Multimedia Information Retrieval, pp. 31–38. ACM (2003)
66. Matthews, R., McDonald, N.J., Hervieux, P., Turner, P.J., Steindorf, M.A.: A wearable physiological sensor suite for unobtrusive monitoring of physiological and cognitive state. In: Proceedings of 29th Annual IEEE International Conference of the Engineering in Medicine and Biology Society, pp. 5276–5281. IEEE (2007)
67. Healey, J.A., Picard, R.W.: Detecting stress during real-world driving tasks using physiological sensors. IEEE Trans. Intell. Transport. Syst. **6**, 156–166 (2005)
68. McDuff, D., Karlson, A., Kapoor, A., Roseway, A., Czerwinski, M.: AffectAura: An intelligent system for emotional memory. In: Konstan, J.A., Chi, E.H., Höök, K. (eds.) Proceedings of the ACM SIGCHI Conference on Human Factors in Computing Systems, pp. 849–858 (2012)
69. Schaaff, K., Degen, R., Adler, N., Adam, M.T.P.: Measuring affect using a standard mouse device. Biomed. Eng./Biomedizinische Technik **57**, 761–764 (2012)

A Refined Examination of Worker Age and Stress: Explaining How, and Why, Older Workers Are Especially Techno-Stressed in the Interruption Age

Stefan Tams

Abstract The workforce is aging rapidly, with the number of older workers increasing sharply (older being defined as 60 and over). At the same time, interruptions mediated by modern information technologies are proliferating in organizations. These interruptions include email notifications and instant messages, amongst others, which have been shown to have hazardous consequences for employees in terms of stress. Older workers might be especially affected by these interruptions, implying major problems for this fast-growing user group with regard to their well-being and work performance. The present study tests a research model suggesting that older workers experience more interruption-based technostress than their younger counterparts because of differences in inhibitory control between older and younger adults. In doing so, this study answers recent calls for examining age as a substantive variable in IS research, and it contributes to the literature on technostress by showing how technostress affects different user groups to different extents.

Keywords Age • Older workers • Inhibition • Stress • Interruptions • Technostress

1 Introduction

The workforce is aging rapidly; in fact, it is said to be *graying* [1–3]. According to the OECD, all of its member countries are experiencing rapid population and workforce aging [2, 4–6]. As workers age, cognitive and biological changes occur. These changes include, for instance, the reduced ability to operate computer input devices, challenges related to memory decline, and the reduced ability to ignore distracting information in the work environment and remain concentrated on the task at hand [7–9]. As a result of these changes, older workers may experience major problems interacting with the ever more modern information technologies at

S. Tams (✉)
Department of Information Technologies, HEC Montréal, Montréal, QC, Canada
e-mail: stefan.tams@hec.ca

F.D. Davis et al. (eds.), *Information Systems and Neuroscience*, Lecture Notes in Information Systems and Organisation 16, DOI 10.1007/978-3-319-41402-7_22

work [10], and they are more at risk of being affected by technological stressors [11].

While the workforce is graying, it also has to deal with an ever-growing number of interruptions mediated by IT, such as a constant stream of instant messages and email notifications. These technology-mediated (T-M) interruptions relentlessly call for workers' attention, causing stress [12–14]. The stress linked to IT, termed *technostress*, can have harmful consequences for workers in the form of elevated levels of stress hormones, causing health problems [15–18]. Organizations are also affected since workers' performance on IT-based tasks suffers under stress [19].

The negative consequences of T-M interruptions might be especially problematic for a graying workforce [19]. Yet, little guidance exists for software designers on how to design IT with older workers in mind [10]. This research investigates the question of *whether, how, and why age impacts physiological stress in an interruption context.*

The next section presents the Inhibitory Deficit Theory of Cognitive Aging to develop a theoretical model of age and technostress. The model explains that age impacts physiological stress in an interruption context due to older peoples' deficits in inhibiting attentional responses to interrupting stimuli. The third section reports on the model testing and results. The paper ends with an account against age discrimination.

2 Background and Hypotheses

This study answers recent calls for more explicit theorizing as to the role of age in IS research [10]. A research agenda paper suggested that IS scholars theorize "touch points of age" [10], which constitute theoretical points through which age touches on various IS phenomena. This notion suggests that age should be modeled as an indirect cause of IS dependent variables, with an added mediator explaining precisely how and why age matters in a given phenomenon. Examining the role of age in IS phenomena through touch points is important so as to avoid justifying it on stereotypical accounts, as commonly done [10]. In fact, using touch points of age can shed ample light on the theoretical nature of the role of age in IS phenomena. In this study, we examine a touch point of age that might be especially relevant in an interruption context: older peoples' deficits in inhibiting attentional responses to interrupting stimuli.

We ground our study and its touch point of age in the literatures on cognition and cognitive aging [20, 21]. These literatures emphasize the importance of working memory to our study context. Working memory is a temporary storage and processing element in the brain that holds the information required for completing an active task [20]. For instance, once a phone number has been looked up, it is held in working memory until it is fully dialed [20]. The capacity of working memory is strictly limited, at times to as few items as one [22]. Given this stringent capacity limitation, interruptions can interfere with task-related information processing when they enter working memory (interrupting stimuli enter working memory when people attend to them) because interruptions draw working memory resources away from the task at hand, leaving fewer resources for task-related information processing. As a result, task-related mental work becomes slow and error-prone (e.g., slow and incorrect dialing of phone numbers or slow reading and writing), leading to stress [11, 20].

To avoid stress by ensuring that only relevant information enters working memory, human cognition relies on selective attention [21]. Selective attention refers to the ability to selectively attend to some information sources while ignoring others; it is concerned with the allocation of processing resources. One important mechanism of selective attention is inhibition [23]. The inhibitory mechanism of selective attention is meant to filter out irrelevant stimuli, helping people remain concentrated on a given active task. Effective inhibition involves actively disregarding any distracting stimuli in a task environment. It is goal-directed; people with effective inhibition can deliberately ignore an interrupting stimulus when they have the goal of remaining concentrated and focused on the task at hand. Yet, the effectiveness of attentional inhibition differs greatly across individuals, most importantly as a function of age [8, 21].

Research on cognitive aging examines the changes in working memory and selective attention that occur as people age [24]. We draw on the inhibitory deficit theory of cognitive aging, which explains how and why age can impact the active control of the contents of working memory [25].[1] This theory is considered a major

[1]Several theories of age differences in memory and attention have been proposed, including the theories of environmental support, deliberate processing, failure to integrate context, working memory demands, perceptual speed, and failures of inhibition [24]. The theory of environmental support suggests that older adults need more support to engage in effective information processing than younger people, implying that they are less efficient at self-initiated information processing. Similar to the preceding theory, deliberate processing implies that the magnitude of age differences in memory depends on the extent to which a tasks involves deliberate information processing instead of more automatic, habitual processing. Failure to integrate context is a concept that is more specific than the preceding one, suggesting that older adults cannot take advantage of contextual cues because they have problems integrating the memory context with the information they are attempting to remember. The theory of working memory demands is simple, implying, on the basis of the large and reliable age differences that have been found in working memory tasks, that older adults' disadvantages in information processing increase with increasing task demands on working memory. Similarly, the theory of perceptual speed indicates that older adults are slower and process information more slowly than their younger counterparts as a function of the cognitive complexity of a task. Finally, the theory of failures of inhibition posits that older adults are much less able than younger people to actively disregard interrupting stimuli in a task environment [24]. In summary, large age differences in memory and attention can be found to the extent to which a task lacks environmental support, requires deliberate processing, involves integrating to-be-remembered information with context, has large working memory demands, can be influenced by perceptual speed, or requires inhibition of irrelevant information [24]. For this study, we deem the theories of environmental support, deliberate processing, and failure to integrate context as not relevant since our focus is on inhibiting attentional responses to interruptions rather than on support or context. Similarly, while the theories of working memory demands and perceptual speed are applicable to our study context, they are too generic to effectively inform

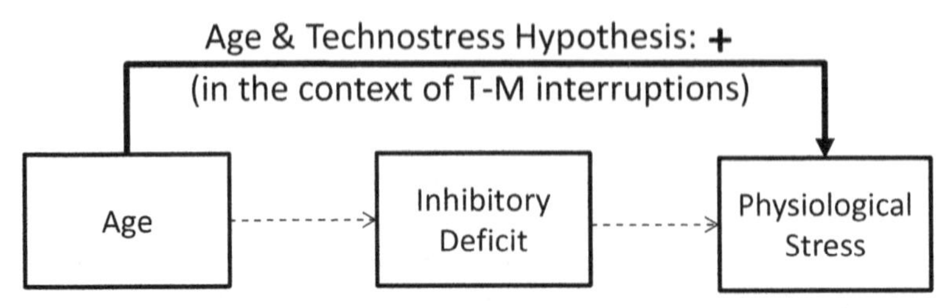

The **line in bold** represents our mediation hypothesis: when encountering T-M interruptions during a task, older people will show higher physiological stress levels than younger people due to an inhibitory deficit (i.e., lower selective attention performance, also understood as less efficient top-down attentional control).

The **dotted lines** represent related direct effects, which are not the focus of our mediation hypothesis but are modeled here only to show what direct effects make up this hypothesis.

Fig. 1 Research model

theoretical approach to aging [26], and it has received widespread empirical support [27–29]. The inhibitory deficit theory holds that attentional inhibition is impaired in older individuals. This impairment is a function of the changes that occur in the frontal lobe, which is the brain component that houses the inhibitory mechanism of selective attention [30, 31]. As people age, the frontal lobe's connections with other brain areas deteriorate, and the frontal lobe shrinks. These anatomical declines in the frontal lobe imply corresponding corrosions in the inhibitory mechanism (because the latter is housed in the frontal lobe). The corrosions in the inhibitory mechanism result in the inhibitory deficit that older individuals commonly demonstrate [31]. As a result, in older compared to younger people interrupting stimuli are more likely to gain access to working memory and to interfere with task-related information processing [25].

In sum, we hold that T-M interruptions can be problematic due to the capacity limitation of working memory. Yet, an effective inhibitory mechanism of selective attention can buffer against the negative consequences of T-M interruptions by preventing them from entering working memory in the first place. This mechanism is a protective shield against interruptions. Still, inhibition is impaired in older adults. We hypothesize that, in the context of T-M interruptions, age is positively associated with physiological stress via increases in inhibitory deficits (i.e., inhibitory deficits mediate the positive correlation between age and physiological stress; see Fig. 1, Table 1).

---

the development of our research hypotheses. The concept of failures of inhibition, however, is directly applicable to our study.

Table 1 Construct definitions

Construct	Definition
Age	Chronologically younger compared to older individuals
Inhibitory deficit	Extent to which a person has a deficit in deliberately suppressing the mental processing of interrupting stimuli in a task environment such that these stimuli do not gain access to mental resources (e.g., working memory resources); a person's inhibitory deficit reflects lower selective attention performance compared to other people
Physiological stress	The extent of physiological strain people experience as a result of their interactions with technologies

3 Methodology and Results

To establish the link between age and stress in people who encounter interruptions, we conducted a correlational study. We recruited younger and older participants and invited them to come to a laboratory, in which they performed a working memory task. While the participants were working on the task, interruptions in the form of instant messages appeared on the screen within specific time intervals. Consistent with past research [32], we instructed the participants to ignore the interruptions. Good task performance was incentivized to increase the relevance of the task for the participants and their involvement. Of the 128 participants we recruited, half represented younger users (age < 30) and half represented older users (age > 60), implying that age was measured as a dichotomous variable with two categories: younger users and older users. This operationalization of age was consistent with research on cognitive aging in general, and with the inhibitory deficit theory in particular [25, 27–29].

Inhibitory Deficit was evaluated using the Stroop color-word task [33]. This task is the most widely-accepted test for inhibition because it most accurately maps onto the definition of inhibitory deficit [34]. The task requires the participants to ignore certain signals that are attentionally compelling but unwanted; the participants have to suppress their attentional responses to these signals while working on a primary task. This procedure is a direct test of someone's ability to inhibit an attentional response to an irrelevant or intrusive piece of information. More specifically, the Stroop task presents to the participants color names that are printed in non-consistent ink colors. The participants must, then, actively inhibit their attentional responses to the printed names of the colors, while selectively attending only to the ink color in which the words are printed. To illustrate, a participant might have to name the ink color *white* for a word that reads *black* (see Fig. 2). Since most people have a natural and strong tendency to read, they must inhibit their attentional response to reading the word *black* in order to correctly name the ink color *white* [34]. This task yields the Stroop effect, which is the difference between participants' response times with congruent and incongruent words. The Stroop task has

Fig. 2 The Stroop color-word task

Instructions: Say the colors of the words on the right while ignoring what the words say. Good luck!

LIGHT GRAY

DARK GRAY

WHITE

BLACK

widespread support in the literature, with test–retest reliabilities ranging from 0.83 to 0.91 [35].[2]

Stress was assessed using salivary α-amylase. A-amylase serves as a cutting-edge marker of the sympathetic nervous system component of the psychobiology of stress by reflecting changes in the stress hormone adrenalin (a particularly relevant hormone in the context of computer-related work) [19, 37]. A-amylase, in and of itself, is a digestive enzyme produced in the salivary glands. As such, it can be found in saliva in relatively high concentrations. As α-amylase is sensitive to alcohol, food, dairy, and caffeine consumption, we controlled for these factors in our analysis. As is necessary for most physiological research, we also controlled for baseline values of α-amylase.

As regards data analysis, we used Preacher and Hayes' [38] standard SPSS macro for testing indirect effects with a 95 % confidence interval and 1000 bootstrap resamples. The results showed that the hypothesized indirect effect was positive as expected and significant. Since zero was outside the 95 % confidence interval, we can conclude with 95 % confidence that the indirect effect of age on participants' physiological stress levels via their inhibitory deficits was different from zero (unstandardized coefficient $b = 1.210$, Std. Error $= 0.818$, $p < 0.05$, $LL = 0.058$, $UL = 3.921$).

4 Discussion and Conclusion

Drawing on the inhibitory deficit theory of cognitive aging, this study confirmed that older users experience more physiological stress in today's technology-driven interruption age than younger ones. The study demonstrated that this effect of age is due to differences in users' abilities to ignore T-M interruptions in a computer-mediated task environment and remain focused on the task at hand. Thus, this study makes an important theoretical contribution to IS research, explaining not only

[2]It is important to note that the Stroop task is not a stress test but a measure of cognitive functioning. In particular, short Stroop tasks like the one employed here (duration of less than ten minutes) cannot be expected to create stress [36]. Further, the Stroop task employed here was not an experimental treatment but simply a measure of inhibitory deficit (a trait).

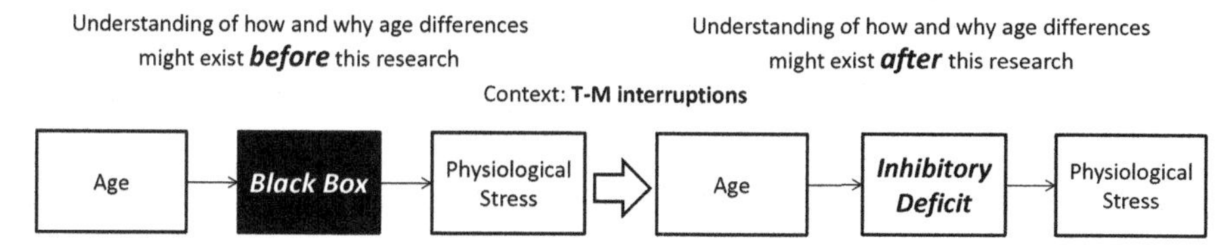

Fig. 3 Theoretical contribution: graying workforces, T-M interruptions, and technostress

whether older users experience more technostress in response to T-M interruptions than younger users but also *how and why* they are more bothered by these interruptions (see Fig. 3).

Given the facts that the workforce is rapidly growing older across all OECD member countries and that modern technologies such as interruptions are proliferating, our findings are alarming. However, the findings should not be construed as encouraging discrimination against older workers or as encouraging any kind of stereotyping. Instead, they should be considered a means of facilitating the development of intervention strategies that can increase the well-being and productivity of a graying workforce. Ultimately, managerial interventions and technological developments are needed that offset the weaknesses and leverage the strengths of older workers to improve their well-being and increase organizational productivity.

Acknowledgements This research was supported by the Fonds de recherche du Québec—Société et culture.

References

1. OECD: Work-force ageing in OECD countries. In: OECD Employment Outlook 1998, pp. 123–151. OECD Publishing, Paris (1998)
2. OECD: Helping older workers find and retain jobs. In: Pensions at a Glance 2011: Retirement-Income Systems in OECD and G20 Countries, pp. 67–79. OECD Publishing, Paris (2011)
3. Pham, S.: The graying work force. The New York Times. http://newoldage.blogs.nytimes.com/2010/11/30/the-graying-workforce/?_php=true&_type=blogs&_r=0 (2010, November 30)
4. OECD: The ageing challenge. In: Ageing and the Public Service: Human Resource Challenges, pp. 17–28. OECD Publishing, Paris (2007)
5. OECD: OECD Employment Outlook 2013. OECD Publishing, Paris (2013)
6. OECD: Editorial. In: International Migration Outlook 2013, pp. 9–10. OECD Publishing, Paris (2013)
7. Charness, N., Holley, P., Feddon, J., Jastrzembski, T.: Light pen use and practice minimize age and hand performance different in pointing tasks. Hum. Factors **46**(3), 373–384 (2004)
8. Darowski, E.S., Helder, E., Zacks, R.T., Hasher, L., Hambrick, D.Z.: Age-related differences in cognition: The role of distraction control. Neuropsychology **22**(5), 638–644 (2008)
9. Salthouse, T.A., Babcock, R.L.: Decomposing adult age differences in working memory. Dev. Psychol. **27**(5), 763–776 (1991)
10. Tams, S., Grover, V., Thatcher, J.: Modern information technology in an old workforce: Toward a strategic research agenda. J Strat Inform Syst **23**(4), 284–304 (2014)

11. Tams, S.: The role of age in technology-induced workplace stress. Unpublished Doctoral Dissertation, Clemson University (2011)
12. Spira, J. B., Feintuch, J. B.: The Cost of Not Paying Attention: How Interruptions Impact Knowledge Worker Productivity. Basex, New York (2005)
13. Tams, S., Thatcher, J., Grover, V., Pak, R.: Selective attention as a protagonist in contemporary workplace stress: Implications for the interruption age. Anxiety Stress Coping **28**(6), 663–686 (2015)
14. Tams, S., Thatcher, J., Ahuja, M.: The impact of interruptions on technology usage: Exploring interdependencies between demands from interruptions, worker control, and role-based stress. In: Davis, F.D. et al. (eds.) Lecture Notes in Information Systems and Organisation: Information Systems and Neuroscience, vol. 10, pp. 19–26. Springer, Heidelberg (2015)
15. Riedl, R., Kindermann, H., Auinger, A., Javor, A.: Technostress from a neurobiological perspective—System breakdown increases the stress hormone cortisol in computer users. Bus. Inform. Syst. Eng **4**(2), 61–69 (2012)
16. Riedl, R.: On the biology of technostress: Literature review and research agenda. Database Adv. Inform. Syst. **44**(1), 18–55 (2013)
17. Riedl, R.: Zum Erkenntnispotenzial der kognitiven Neurowissenschaften für die Wirtschaftsinformatik: Überlegungen anhand exemplarischer Anwendungen. NeuroPsychoEconomics **4**(1), 32–44 (2009)
18. Galluch, P.S., Grover, V., Thatcher, J.B.: Interrupting the workplace: Examining stressors in an information technology context. J Assoc Inform Syst **16**(1), 1–47 (2015)
19. Tams, S., Hill, K., Ortiz de Guinea, A., Thatcher, J., Grover, V.: NeuroIS—Alternative or complement to existing methods? Illustrating the holistic effects of neuroscience and self-reported data in the context of technostress research. J Assoc Inform Syst **15**(10), 723–753 (2014). Article 1
20. Wickens, C.D., Lee, J., Liu, Y.D., Becker, S.G.: Introduction to Human Factors Engineering. Prentice Hall, Upper Saddle River, NJ (2004)
21. Strayer, D.L., Drews, F.A.: Attention. In: Perfect, T.J. (ed.) Handbook of Applied Cognition, 2nd edn, pp. 29–54. Wiley, Hoboken, NJ (2007)
22. Dumas, J.A., Hartman, M.: Adult age differences in the access and deletion functions of inhibition. Aging Neuropsychol Cognit **15**(3), 330–357 (2008)
23. Houghton, G., Tipper, S.P.: A model of inhibitory mechanisms in selective attention. In: Carr, T.H. (ed.) Inhibitory Processes in Attention, Memory, and Language, pp. 53–112. Academic, San Diego, CA (1994)
24. Smith, A.D.: Memory. In: Salthouse, T.A. (ed.) Handbook of the Psychology of Aging, 4th edn, pp. 236–250. Academic, San Diego, CA (1996)
25. Hasher, L., Zacks, R.T.: Working memory, comprehension, and aging: A review and a new view. In: Bower, G.H. (ed.) The Psychology of Learning and Motivation: Advances in Research and Theory, vol. 22, pp. 193–225. Academic, San Diego, CA (1988)
26. McDowd, J.M., Shaw, R.J.: Attention and aging: A functional perspective. In: Salthouse, T.A. (ed.) The Handbook of Aging and Cognition, 2nd edn, pp. 221–292. Lawrence Erlbaum Associates, Mahwah, NJ (2000)
27. Carlson, M.C., Hasher, L., Connelly, S.L., Zacks, R.T.: Aging, distraction, and the benefits of predictable location. Psychol Aging **10**(3), 427–436 (1995)
28. Hasher, L., Stoltzfus, E.R., Zacks, R.T., Rypma, B.: Age and inhibition. J Exp Psychol Learn Mem Cognit **17**(1), 163–169 (1991)
29. Zacks, R., Hasher, L.: Cognitive gerontology and attentional inhibition: A reply to burke and McDowd. J Gerontol B Psychol Sci Soc Sci **52B**(6), P274 (1997)
30. Luria, A.R.: The Working Brain: An Introduction to Neuropsychology. Basic, New York (1973)
31. Jurado, M., Rosselli, M.: The elusive nature of executive functions: A review of our current understanding. Neuropsychol Rev **17**(3), 213–233 (2007)

32. Theeuwes, J.: Cross-dimensional perceptual selectivity. Percept Psychophys **50**(2), 184–193 (1991)
33. Stroop, J.R.: Studies of interference in serial verbal reactions. J Exp Psychol **18**(6), 643–662 (1935)
34. Shilling, V.M., Chetwynd, A., Rabbitt, P.M.A.: Individual inconsistency across measures of inhibition: An investigation of the construct validity of inhibition in older adults. Neuropsychologia **40**(6), 605 (2002)
35. Spreen, O., Strauss, E.: A Compendium of Neuropsychological Tests. Oxford University Press, New York (1998)
36. Renaud, P., Blondin, J.P.: The stress of Stroop performance: Physiological and emotional responses to color–word interference, task pacing, and pacing speed. Int J Psychophysiol **27** (2), 87–97 (1997)
37. Korunka, C., Huemer, K.H., Litschauer, B., Karetta, B., Kafka-Lützow, A.: Working with new technologies: Hormone excretion as an indicator for sustained arousal. A pilot study. Biol Psychol **42**(3), 439–452 (1996)
38. Preacher, K.J., Hayes, A.F.: SPSS and SAS procedures for estimating indirect effects in simple mediation models. Behav Res Meth Instrum Comput **36**(4), 717–731 (2004)

Consumer Grade Brain-Computer Interfaces: An Entry Path into NeuroIS Domains

Nash (Nebojsa) Milic

Abstract This research provides a high-level and non-technical insight into the current state of NeuroIS field and a concise overview of consumer grade Brain-Computer interfaces (BCI). Special attention is given to the electroencephalography (EEG) based BCIs. Additionally, a two-dimensional overview with five clusters (Surgical, Clinical, Lab, Office and Plug & play cluster) is proposed as a mean to position NeuroIS tools based on the resource requirements and ease of use. Finally, this research employs a consumer grade BCI (Neurosky Mindwave) in a pilot study and argues for future research on consumer grade BCIs for topics related to information overload.

Keywords BCI • EEG • Brain-computer interfaces • Consumerization • Consumer grade • Plug & play • Information overload

1 Introduction

This research understands a brain-computer interface (BCI) as a sum of hardware and software elements of a communications system that permits cerebral activity to control computers or external devices [1–3]. Consumer grade BCIs represent a special BCI subgroup targeted at consumer market. Consumer BCIs are readily available, easy to install and comfortable to use.

The main goal of this research is to position consumer grade BCIs among other commonly used NeuroIS tools and to test the technical capabilities of a specific consumer grade BCI (EEG based Neurosky Mindwave) in a pilot study. This research proceeds as follows: a brief literature review of thematic, methodological and theoretical advances in NeuroIS field is provided; followed by a two dimensional comparison of brain-computer interfaces with widely used NeuroIS tools and with a set of available consumer grade brain-computer interfaces. This research then presents the results of a pilot study of a consumer grade BCI, namely Neurosky

N.(N.) Milic (✉)
Baylor University, Waco, TX, USA
e-mail: nash_milic@baylor.edu

F.D. Davis et al. (eds.), *Information Systems and Neuroscience*, Lecture Notes in Information Systems and Organisation 16, DOI 10.1007/978-3-319-41402-7_23

Mindwave, and concludes with suggestions for future consumer grade BCI research.

2 Literature Review

In recent years, the Information Systems (IS) discipline has begun to embrace neural instruments and methods with the intention of complementing existing topics and expanding the understanding of the emergent phenomena. This has led to the creation of a specific branch of IS research, known as Neuro-Information Systems (NeuroIS). NeuroIS has improved the capacity of IS research significantly and a number of research breakthroughs have occurred as a result. NeuroIS has recently been used to answer multiple questions, ranging from understanding technostress [4], information processing biases in virtual teams [5], understanding effects of emotional states on financial trading decisions [6], using measures of risk perception to predict information security behavior [7] and complementing business process modeling tools [8] to name a few.

Thus, it is not surprising that IS scholars started to lay theoretical and methodological grounds on which future research can be built. Reflections on the Gmunden Retreat from 2009 [9] re-defined what NeuroIS is, which tools are relevant for NeuroIS, what IS can learn from neuroscience and what were the current challenges for NeuroIS at that time. A research commentary [10] illustrated the potential of cognitive neuroscience for IS research, especially when it comes to localizing the neural correlates of IS constructs, capturing hidden mental processes and challenging assumptions and enhancing IS theories. Vom Brocke and Liang [11] also contributed to the discipline with a set of guidelines for NeuroIS studies. Those guidelines are designed to help researchers better understand phases typical for NeuroIS research and to guide NeuroIS research through the emerging standards of the discipline. Tams et al. [4] used a technostress study to illustrate the holistic effects that come from using neurosciences and self reported data in tandem. Tams et al. improved our understanding on triangulating different sources of quantitative data by showing the scenario in which different measures can constitute alternatives and/or complements in the prediction of theoretically-related outcomes. Specifically, Tams et al. demonstrated that physiological and psychological measures can actually lead to divergent findings. Furthermore, Gregor et al. [12] developed a nomological network with an overarching view of relationships among emotions and common constructs of interest in NeuroIS research. Finally, Müller-Putz et al. [13] ventured deeply into the foundations, measurements and application of electroencephalography in IS. By publishing their work, Müller-Putz et al. equipped prospective NeuroIS researchers with solid methodological foundations for conducting EEG based research.

3 NeuroIS Tools: Clinical and Consumer Grade Devices

Although NeuroIS continues to prove its value by expanding the knowledge on multiple IS related phenomena; and although sound theoretical and methodological foundations of NeuroIS have been laid out, many IS researchers can still feel reluctant to venture into NeuroIS topics. Some are turned away by significant resources required to conduct a NeuroIS study [14], while others are discouraged by the sheer breadth of non-IS and highly technical knowledge required to successfully conduct a state of the art NeuroIS experiment [13]. In general, NeuroIS research requires researchers to select a proper instrument and equipment that will adequately detect all elicited aspects of researched phenomena, create and maintain all required parameters for the instrument to operate optimally and finally to analyze the readings from the clinical grade neural interfaces [4, 11, 13]. Few IS researchers are properly trained to conduct those studies and some can be deterred by administrative hurdles required to conduct medical-grade research on human subjects. All those factors combined prevent NeuroIS from becoming one of the dominant areas of IS research.

Additionally, a preconceived notion of complexity, ambiguity and dangers [15, 16] of neural instruments may also be one of the culprits for a relatively low proliferation of neural technologies outside the academia. Specifically, it has been known for four decades that the human brain can communicate directly to computers via brain-computer interfaces [1], yet proliferation of those interfaces never happened. Companies like Emotiv, Neurosky and Microsoft are known to work on BCIs [9] and multiple BCIs are even available to a wide consumer audience—but to no avail. Organizations and individual users are still reluctant to include BCIs into their IS infrastructure.

That situation on the ground can perplex IS researchers: Why is it that despite growing research momentum, wide use of BCI did not happen? This research posits that a wide proliferation of BCIs did not happen because the consumer side of NeuroIS is still largely unexplored. In order to explain our reasoning, an overview of the most common NeuroIS tools with accompanying acronyms is presented below (Fig. 1). In this figure, the most commonly used NeuroIS tools are classified based on two dimensions: Ease of use (x axis) and Resource requirements (y axis). Ease of use is to be understood as the level of efforts required to use a NeuroIS tool: on the one hand, fMRI requires extensive efforts to set up and run, and considerable efforts to process the outputs of the device. On the other hand, a SCR requires significantly lower amount of efforts, since its use is much more natural to both experimenters and participants. The resource requirements scale should be perceived as the amount of resources (i.e. financial resources, infrastructure, manpower and time) required to successfully use a NeuroIS tools. For the purposes of illustration let us compare a PET and SCR on this scale: a PET requires a clinical level of infrastructure and comes with large upfront and running costs, which

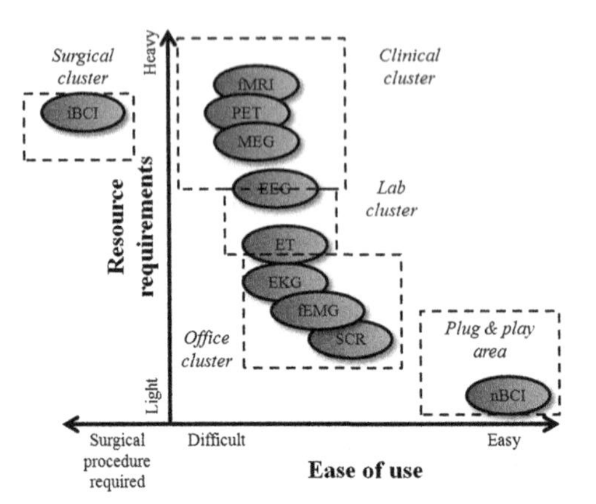

LEGEND:

iBCI- Invasive Brain-Computer Interfaces
fMRI - Functional Magnetic Resonance Imaging
PET - Positron Emission Topography
ET - Eye Tracking
SCR - Skin Conductance Response
EKG - Electrocardiogram
fEMG - Facial Electromyography
EEG - Electroencephalography
MEG - Magnetoencephalography
nBCI - noninvasive Brain-Computer Interfaces

Fig. 1 An overview of NeuroIS tools

dwarfs the resource requirements for SCR.[1] The listed NeuroIS tools in Fig. 1 are grouped together into clusters: Clinical cluster represent those devices which are mostly employed in a clinical settings; Lab cluster is positioned between the Clinical and Office cluster since most of the devices that populate this cluster are commonly used in laboratory settings; Office cluster consists of those devices that can be easily used in regular office settings. Surgical cluster is populated with invasive BCIs which tend to stay outside the reach of non medical disciplines [15]. Finally, the Plug & play area depicts all NeuroIS tools which require little to no special conditions and which are ready to use with little or no difficulties.

Clinical, Lab and Office clusters have been extensively used within the NeuroIS literature and the available body of knowledge that originates from those clusters is expanding. However, the Plug & play area represents uncharted waters for NeuroIS and for potential organizational and individual users of BCIs. Noninvasive BCIs are cheap and easy to use, but little is known about their research or usage potentials. Apart from few pioneering studies (e.g. [5, 17]) NeuroIS has not employed consumer grade noninvasive BCI extensively.

In order to further explore the Plug and play cluster, which is exclusively populated by noninvasive BCI devices, this research provides a concise and non exhaustive[2] overview of different consumer grade BCIs. According to their respective manufacturers, those devices are able to compliment NeuroIS research and organizational needs at relatively low resource demands, while being easy to use. One device from this cluster (Emotiv Epoc) was used in the previously mentioned studies [5, 17].

[1]Dimensions of "Resource requirements" and "Ease of use" were adapted from Riedl et al. [9] and Dimoka et al. [14].

[2]Please consult [18] for a more exhaustive overview of wearable EEG sensors and interfaces.

4 Mindwave Neurosky Pilot Experiment

With that in mind and with an intention to test a consumer level BCI, a pilot study with one of the devices from Table 1 was conducted. The selected device, Neurosky Mindwave headband (highlighted in Table 1; headband in the following text), was used to gather data from a group of 12 test subjects.[3] Neuro Experimenter software v3.28 was used to access the API of the headband and to record EEG based BCI data [20]. All participating subjects of this pilot study were healthy PhD students at a medium-sized private university in the southern part of the United States. All subjects were right handed, and between the ages of 27 and 35. Two participants were female. All recordings were gathered in a standard office environment, while subjects were working on light office tasks that required them to use a computer (e.g. checking email inbox, browsing the Internet and arranging files etc.).

According to software manufacturer's specification, the headband and the software used in this pilot study records brainwave readings every 500 ms, via a "cluster sensor" positioned on the participant's forehead and targeted at the prefrontal cortex (PFC). PFC is known to be the executive center of the human brain [10, 14, 21], where decision actions (e.g. calculations) are performed. Descriptive statistics of the gathered data are presented in Table 2. In depth analysis of gathered data (e.g. outliers and ERPs) was omitted as a result of technical constraints of this venue of publication.

To sum, this pilot study continues a research direction set by previous studies [21, 22] by demonstrating that a consumer grade BCI recording only one single channel (Fp1) can be used to collect EEG signals in a regular office environment. Naturally, spatial and temporal resolutions of the recordings are not identical to the standards that are generally employed in clinical level EEG studies [13, 21]—instead of dozens of simple sensors, the headband used in this pilot study has only one "clustered sensor"; and instead of the recommended 200 ms temporal resolution this system (i.e. the headband and the employed software) is capable of recording only 500 ms intervals. However, according to criteria presented in Kübler et al. 2001 [1], the headband used in this study fits all the requirements for a BCI because it successfully detected the electrophysiological activity of the user's brain, recorded the signals at 97.62 % accuracy (which is above the proposed 70 % threshold) and bypassed most of the stated limitations.

[3]All recorded data is available for download from here: https://goo.gl/bOcDGt.

Table 1 A non-exhaustive overview of consumer grade BCIs

Name	EEG	EMG	ECG	EOG	Manufacturer	Released
Open BCI Ganglion (R&D)	×	×	×		Open BCI	Summer 2016
iFocusBand	×				iFocus	October 2014
Emotiv Epoc	×				EmotivSystems	December 2009
Emotiv Insight	×				EmotivSystems	August 2015
Muse	×				InteraXon	April 2014
Aurora Headband				×	Iwinks	July 2015
MindSet	×				NeuroSky	March 2007
MyndPlay	×				MyndPlay	December 2011
MelonHeadband	×				Melon	Nov 2014
XWave Sonic	×				PLX Devices	February 2013
MindWave	×				NeuroSky	March 2011
Mindflex	×				NeuroSky (Mattel)	December 2009
Nerural Impulse Actuator		×			OCZ	April 2008

EEG Electroencephalogram, *EMG* Electromyograph, *ECG* Electrocardiogram, *EOG* Electrooculograp; adapted from [19]

Table 2 Descriptive statistics of the data collected from Neurosky Mindwave headband

EEG Bands	N	Minimum	Maximum	Mean	Std. deviation	Variance
Delta	16217	0.50	98.64	47.48	27.12	735.28
Theta	16217	0.17	81.30	21.22	13.01	169.25
Alpha 1	16217	0.01	71.88	7.45	6.93	47.98
Alpha 2	16217	0.04	58.22	6.60	6.12	37.45
Beta 1	16217	0.06	45.21	6.08	5.57	31.02
Beta 2	16217	0.04	46.94	5.58	5.09	25.89
Gamma 1	16217	0.02	37.46	3.26	3.25	10.54
Gamma 2	16217	0.01	30.44	2.33	2.41	5.83

Number of observations—16217, blinks—2551, Data points with an error/missing recording—394

5 Future Research: How Can Consumer Grade BCIs Help in Detecting Information Overload?"

This pilot study paves the way for using BCIs in NeuroIS to better understand and detects one of the growing technostress phenomena known as information overload (IO). IO can be defined from two perspectives. On the one side, IO occurs on organizational levels when the amount of input to a system exceeds its processing capacity [23] and thus causes negative or unwanted effects for the organization and its employees. On the other side, individuals experience IO as state in which a vast amount of information is readily available, almost instantaneously, without mechanisms to check the validity of the content and the potential risk of misinformation [24, 25]. In short, IO can be understood as a state in which there is just too much information to cope with [26]. Thus, it follows that IO puts a tremendous strain on individuals and on individual cognitive performances—especially on the short term memory function of the human brain and on the executive center of the brain located in the pre frontal lobe [10, 27–29].

In previous research, a set of clinical grade instruments, such as FMR scan or EEG were used to further the understanding of the human brain [10]. However, IO would be difficult to induce in a clinical/lab environment and potential results might be unusable for practitioners (i.e. it is not practical to regularly monitor employees with a FMR scanner). Therefore, if we are to detect and treat IO in real-life organizational scenarios a more mobile and wearable technology is needed. Consumer grade BCIs seems to fit into that description perfectly. With that in mind, a study with two different consumer grade neural interfaces (Epoc Emotiv and Neurosky Mindwave) and real-life tasks is proposed. Ideally, the author hopes that the mentioned study will provide additional insights about detection and prevention of IO overload through the use of BCIs.

6 Contributions

This research brings multiple contributions. Firstly, it provides a concise and non-technical insight into the current state of the NeuroIS field. Secondly, a classification of commonly used neural tools is proposed (Fig. 1). In it, different neural tools are grouped on the basis of resource intensity and ease of use, which lead to the creation of five distinct clusters. Thirdly, special attention has been given to one cluster which was not used as widely as the other clusters and an overview of different noninvasive consumer grade BCIs has been provided (Table 1). Fourthly, this research conducted a preliminary study in which a BCI system is used to collect and process brainwave readings. And finally, this research proposed a potential study in which BCIs can be used to detect, prevent and better understand a rampant problem of information overload.

7 Conclusion

In conclusion, this research might on the one hand encourage the novice and prospective IS researchers to consider joining the growing NeuroIS community by adding neural tools from the resource light and easy to use Plug & play cluster into their research instrumentarium. That should allow novice researchers to circumvent seemingly overwhelming technical and medical demands imposed by the sheer complexity of neuroscience. On the other hand, I hope that this research might point the attention of NeuroIS veterans and practitioners to a promising and partially neglected phenomenon of noninvasive consumer grade BCIs.

References

1. Kübler, A., Kotchoubey, B., Kaiser, J., Wolpaw, J.R., Birbaumer, N.: Brain-computer communication: Unlocking the locked in. Psychol. Bull. **127**, 358–375 (2001)
2. Lotte, F., Congedo, M., Lécuyer, A., Lamarche, F., Arnaldi, B.: A review of classification algorithms for EEG-based brain-computer interfaces. J. Neural Eng. **4**, R1–R13 (2007)
3. Nicolas-Alonso, L.F., Gomez-Gil, J.: Brain computer interfaces, a review. Sensors **12**, 1211–1279 (2012)
4. Tams, S., Hill, K., Ortiz de Guinea, A., Thatcher, J., Grover, V.: NeuroIS—Alternative or complement to existing methods? Illustrating the holistic effects of neuroscience and self-reported data in the context of technostress research. J. Assoc. Inf. Syst. **15**, 723–753 (2014)
5. Minas, R.K., Potter, R.F., Dennis, A.R., Bartelt, V., Bae, S.: Putting on the thinking cap: Using NeuroIS to understand information processing biases in virtual teams. J. Manag. Inf. Syst. **30**, 49–82 (2014)
6. Astor, P.J., Adam, M.T.P., Jerčić, P., Schaaff, K., Weinhardt, C.: Integrating biosignals into information systems: A NeuroIS tool for improving emotion regulation. J. Manag. Inf. Syst. **30**, 247–278 (2013)
7. Vance, A., Anderson, B.B., Kirwan, C.B., Eargle, D.: Using measures of risk perception to predict information security behavior: insights from electroencephalography (EEG). J. Assoc. Inf. Syst. **15**, 679 (2014)
8. Shitkova, M., Holler, J., Jörg, B., Léger, P.-M.: The potentials of neuroscience methods for business process modeling tools. https://www.wi.uni-muenster.de/research/publications/98826
9. Riedl, R., Banker, R., Benbasat, I., Davis, F., Dennis, A., Dimoka, A., Gefen, D., Gupta, A., Ischebeck, A., Kenning, P., Müller-Putz, G., Pavlou, P., Straub, D., vom Brocke, J., Weber, B.: On the foundations of NeuroIS: Reflections on the Gmunden retreat 2009. Commun. Assoc. Inf. Syst. **27**, 243–264 (2010)
10. Dimoka, A., Pavlou, P.A., Davis, F.D.: Research commentary—NeuroIS: The potential of cognitive neuroscience for information systems research. Inf. Syst. Res. **22**, 687–702 (2010)
11. vom Brocke, J., Liang, T.-P.: Guidelines for neuroscience studies in information systems research. J. Manag. Inf. Syst. **30**, 211–234 (2014)
12. Gregor, S., Lin, A.C.H., Gedeon, T., Riaz, A., Zhu, D.: Neuroscience and a nomological network for the understanding and assessment of emotions in information systems research. J. Manag. Inf. Syst. **30**, 13–48 (2014)
13. Müller-Putz, G., Riedl, R., Wriessnegger, S.C.: Electroencephalography (EEG) as a research tool in the information systems discipline: Foundations, measurement, and applications. Commun. Assoc. Inf. Syst. **37**, 911–948 (2015)
14. Dimoka, A., Banker, R.D., Benbasat, I., Davis, F.D., Dennis, A.R., Gefen, D., Gupta, A., Ischebeck, A., Kenning, P., Pavlou, P.A., Müller-Putz, G., Riedl, R., Jan Brocke, V., Weber, B.: On the use of neurophysiological tools in IS research: Developing a research agenda for NeuroIS. MIS Q. **36**, 679–702 (2012)
15. Engber, D.: The neurologist who hacked his brain—And almost lost his mind, http://www.wired.com/2016/01/phil-kennedy-mind-control-computer/
16. Piore, A.: To study the brain, a doctor puts himself under the knife, https://www.technologyreview.com/s/543246/to-study-the-brain-a-doctor-puts-himself-under-the-knife/

17. Kuan, K.K.Y., Zhong, Y., Chau, P.Y.K.: Informational and normative social influence in group-buying: Evidence from self-reported and EEG Data. J. Manag. Inf. Syst. **30**, 151–178 (2014)
18. Balanou, E., van Gils, M., Vanhala, T.: State-of-the-art of wearable EEG for personalized health applications. Stud. Health Technol. Inform. **189**, 119–124 (2013)
19. Comparison of consumer brain–computer interfaces. https://en.wikipedia.org/w/index.php?title=Comparison_of_consumer_brain%E2%80%93computer_interfaces&oldid=700981088 (2016)
20. NeuroExperimenter. http://store.neurosky.com/products/neuroexperimenter
21. Das, R., Chatterjee, D., Das, D., Sinharay, A., Sinha, A.: Cognitive load measurement—A methodology to compare low cost commercial EEG devices. In: 2014 International Conference on Advances in Computing, Communications and Informatics (ICACCI), pp. 1188–1194 (2014)
22. Grierson, M., Kiefer, C.: Better brain interfacing for the masses: Progress in event-related potential detection using commercial brain computer interfaces. In: CHI '11 Extended Abstracts on Human Factors in Computing Systems, pp. 1681–1686. ACM, New York (2011)
23. Speier, C., Valacich, J.S., Vessey, I.: The influence of task interruption on individual decision making: An information overload perspective. Decis. Sci. **30**, 337–360 (1999)
24. Flew, T.: New Media: An Introduction. OUP, South Melbourne (2007)
25. Graham, G.: The Internet: A Philosophical Inquiry. Routledge, London (1999)
26. Sobotta, N.: A systematic literature review on the relation of information technology and information overload. Presented at the Hawaii International Conference on System Sciences (2016)
27. Dimoka, A.: How to conduct a functional magnetic resonance (fMRI) study in social science research. MIS Q. **36**, 811–840 (2012)
28. Fuster, J.M., Bodner, M., Kroger, J.K.: Cross-modal and cross-temporal association in neurons of frontal cortex. Nature **405**, 347–351 (2000)
29. Goldman-Rakic, P.S., Cools, A.R., Srivastava, K.: The prefrontal landscape: Implications of functional architecture for understanding human mentation and the central executive [and discussion]. Philos. Trans. Biol. Sci. **351**, 1445–1453 (1996)

Measuring the Popularity of Research in Neuroscience Information Systems (NeuroIS)

Naveen Zehra Quazilbash and Zaheeruddin Asif

Abstract Research in Neuroscience Information Systems aka NeuroIS has gained considerable momentum in the past decade. The purpose of this study is to examine the landscape of research conducted in the emerging discipline of NeuroIS in order to track the progression in the field. Our analysis is based on a systematic review of 59 papers published in high impact conferences and journals over the past 5 years (2011 to 2015 inclusive). It offers a bird eye view to the upcoming researchers who intend to pursue NeuroIS research.

Keywords NeuroIS literature review • Topical analysis

1 Introduction

Information Systems (IS) studies typically rely on traditional methods of research, primarily surveys and experimental setups. Recently researchers have drawn interest in Cognitive Neuroscience (CN) discipline to address the very issue of decision making (DM) in IS users. The issue of DM has been researched not only in neurosciences but also in IS [1,2]. Many popular IS theories like theory of reasoned action, Technology Acceptance Model, and effort accuracy framework have been used for the said purpose [3]. The IS community is witnessing a growing interest in applying theories, methods and tools in neuroscience to inform IS research [4]. For this purpose, the use of brain imaging tools in the CN discipline has provided a useful model.

This phenomenon has given rise to the sub-field of "NeuroIS", which has consistently gained momentum over the past 5 years. This sub-field employs brain imaging tools to inform its research. Most popular tools in this regard are fMRI, EKG, and EEG. These CN tools can be used along with traditional IS tools like questionnaires, and surveys. Also, triangulation of measures provides greater

---

The term NeuroIS was first coined by Angelika Dimoka in the year 2007.

N.Z. Quazilbash (✉) • Z. Asif
Faculty of Computer Science, Institute of Business Administration, Karachi, Pakistan
e-mail: nquazilbash@iba.edu.pk; zasif@iba.edu.pk

F.D. Davis et al. (eds.), *Information Systems and Neuroscience*, Lecture Notes in Information Systems and Organisation 16, DOI 10.1007/978-3-319-41402-7_24

validation across measures thus enhancing the ecological validity of IS studies [4]. A dedicated conference on the subfield of "NeuroIS" has been held on an annual basis for the last several years. It is called the "Gmunden Retreat on NeuroIS".

Given this interest in this sub-field of IS, the goal of this work is to review and report publications in this area over the past 5 years. This essentially establishes a strong ground for its acceptability in the IS research community and provides a consolidated overview of avenues where aspiring NeuroIS researchers can aim to publish their work.

The rest of the paper is organized as follows. In the next section we discuss the research methodology and study overview. Results are mentioned in Results section. Discussion section presents a discussion of the results and we conclude with a summary in the end.

2 Study Overview

This paper systematically reviews NeuroIS literature being published in high-impact journals (AIS approved "Basket of 8") and conferences (recognized by AIS) during the period from 2011 to 2015. Table 1 list sources used to elicit the articles and their type i.e. journal or conference proceedings. The general purpose of this paper is to integrate the literature published in this domain in the past 5 years. What enhances the rigor of this study is its validity and reliability [5]. Validity in the sense that the databases covered are the registered ones provided on the Association for Information System's (AIS) website. Moreover, the keywords pertain to the exact terms being used in such studies, whereas, the period covered is essentially the one that contains a lower bound from where the specific studies actually started to get published [6] and until the last year. It is reliable in the sense that it documents the above mentioned procedure in enough detail to be replicated. Accordingly, the directorial research question of this study is, *"Whether the NeuroIS subfield of IS has proven its mark as an established and viable area of research in Information Systems?"* The keyword used for this search was "neurois". The reason behind using this keyword only is the idea that if there is an article that necessarily reports this particular kind of research then it will, somewhere in the article, mention this term. This is so because "neurois" is the only keyword that defines work in this field. Moreover, the keywords were searched on the AIS website (http://aisel.aisnet.org/), therefore the results included the related terms (for e.g. fMRI, HCI, trust, brain, emotions etc.) automatically. Table 2 lists down a breakup of total number of publications considered from each source (as listed in Table 1).

3 Results

See Tables 3 and 4 and Figs. 1 and 2.

Table 1 Considered publication outlets

Name	Type
European Journal of Information Systems (EJIS)	Journal
Information Systems Journal (ISJ)	Journal
Information Systems Research (ISR)	Journal
Journal of Association of Information Systems (JAIS)	Journal
Journal of Information Technology (JIT)	Journal
Journal of Management Information Systems (JMIS)	Journal
Journal of Strategic Information Systems (JSIS)	Journal
Management Information Systems Quarterly (MISQ)	Journal
International Conference on Information Systems (ICIS)	Proceedings
Americas Conference on Information Systems (AMCIS)	Proceedings
Pacific Asia Conference on Information Systems (PACIS)	Proceedings
Mediterranean Conference on Information Systems (MCIS)	Proceedings
European Conference on Information Systems (ECIS)	Proceedings

Table 2 Frequency of NeuroIS publications in AIS recognized journals and conferences

Year	2011	2012	2013	2014	2015	Others	Comments
EJIS	0	0	0	0	0	1	2015 online publication; yet to be included in forthcoming issues
ISJ	0	0	0	0	1	–	Only 1 editorial in 2015
ISR	1	0	0	0	0	–	–
JAIS	0	1	0	4	1	–	2014 special issue
JIT	0	0	0	0	0	–	–
JMIS	0	0	1	7	1	–	–
JSIS	0	0	0	0	0	–	–
MISQ	0	2	1	0	0	–	–
Total	1	3	2	11	3	1	Total = 21
ICIS	3	4	3	5	4	–	–
AMCIS	1	0	0	2	7	–	–
PACIS	0	0	0	1	0	–	–
MCIS	0	1	0	0	0	–	–
ECIS	1	1	1	1	3	–	–
Total	5	6	4	9	14	–	Total = 38
Total	6	9	6	20	17	–	21 + 38 = 59

Table 3 Summary of NeuroIS topics and respective methods and tools

Topic	Methods and Tools
EJIS	
Forthcoming	
User's perception and response to security messages. [C] 5	ET
ISJ	
2015	
Editorial on exploring potential of NeuroIS in IS research [O] 2	Editorial
ISR	
2011	
Exploring potential of NeuroIS in IS research [O] 3	Research commentary
JAIS	
2012	
Historical analysis [O] 3	Review paper
2014	
NeuroIS research methodology [O] 3	Methodological paper
NeuroIS for reporting Technostress studies [O] 5	Methodological paper
Users' disregard of security warnings [D] 4	EEG
Application of EFRP to IS research [O] 7	Methodological paper
2015	
Emotions(arousal) and bidding behavior in e-auctions [E] 3	SCR, HR
JMIS	
2013	
Emotional regulation in financial decisions [E, D] 5	Conceptual and design paper
2014	
Understanding and assessment of emotions in IS [E] 5	EEG
Understanding information processing bias [D] 5	EEG
Trust and mentalizing in human-avatar interaction [S] 5	fMRI
User engagement in online gaming [C] 4	EEG
Social influence in group-buying [S] 3	EEG
Perceived ease of use and usefulness [C] 3	EEG
Guidelines for neuroscience studies in IS research [O] 2	Conceptual paper
2015	
Role of self-control in information security violations [C] 3	EEG
MISQ	
2012	
Use of Neurophysiological tools for developing NeuroIS agenda [O] 14	Conceptual and review paper
How to conduct fMRI studies [O] 1	Methodological paper
2013	
IS research and behavioral economics [O] 1	Editorial
ICIS	
2011	

(continued)

Table 3 (continued)

Topic	Methods and Tools
Emotions (arousal) in electronic markets (auctions) [E] 4	Methodological paper
Online payment method choice [D] 2	ET
Trust and mentalizing in human-avatar interaction [S] 5	fMRI
2012	
Consumer's online cognitive scripts [C] 4	EEG
Connectivity analysis in NeuroIS research [O] 5	Methodological paper
Emotions in online financial markets wih humans and agents [E] 3	Conceptual paper
Startle reflex modulation for measuring affective information processing [O] 2	Methodological paper
2013	
Effect of arousal and valence on mouse interaction [E] 3	Mouse cursor movement
Trust in social networking sites [S] 5	fMRI
Experience and pressure in IS usage behavior [C] 7	ET
2014	
Habituation to security warnings [C] 5	fMRI
Consumer impulsiveness in online trust evaluations [S] 4	fMRI
Variance and process types of IS research [O] 2	Methodological paper
Emotions and Aesthetics in ICT Usage [E] 2	EEG
Content familiarity in web design [E] 1	Conceptual paper
2015	
Emotion, attention and behavioral control [E] 2	fEMG
System un-related emotions in adoption of IS [E] 2	Conceptual paper
Information avoidance behavior [C] 1	Conceptual paper
Enterprise Resource Planning (ERP) system design [E] 1	Conceptual/behavioral paper
AMCIS	
2011	
NeuroIS research methods [O] 2	Review paper
2014	
Cognition in online shopping markets [C] 2	fMRI
Trust [S] 3	Conceptual paper
2015	
Privacy paradox [D] 2	Conceptual paper
Technostress and users' response to security warnings [E] 5	Hormone (saliv. str.)
Human-Information-Processor Model View of e-Government [C] 2	Conceptual paper
Organizational Memory [C] 2	Conceptual paper
TAM [C] 1	Conceptual paper
Technostress [O] 1	Methodological paper
Visual aesthetics [E] 4	GSC, fEMG, behavioral paper

(continued)

Table 3 (continued)

Topic	Methods and Tools
PACIS	
2014	
Metacognition in e-commerce [C] 2	fMRI
MCIS	
2012	
Technology acceptance [C] 5	EEG
ECIS	
2011	
Emotions (regret) in electronic markets (auctions) [E] 4	SC
2012	
Human centered design [O] 3	Methodological paper
2013	
Emotional regulation in financial decisions [E, D] 3	Conceptual and design paper
2014	
Emotions (arousal) in electronic markets (auctions) [E] 3	EEG, SCR, HR
2015	
Emotional regulation in financial decisions [O] 4	Methodological paper
Job Strain Information System Service (JSISS) [E] 9	Conceptual and design paper
Technostress [O] 3	Methodological and design paper

Acronyms: *EEG* electroencephlography, *ET* eye-tracking, *fEMG* facial electromyography, *fMRI* functional magnetic resonance imaging, *GSC* Galvanic skin conductance, *HR* heart rate, *saliv. str.* salivary stress, *SC* skin conductance, *SCR* skin conductance response. Code for paper categorization: [C] = cognitive processes, [E] = emotional processes, [S] = social processes and [D] = decision-making processes. The code is followed by the number of authors of the papers

Table 4 Methods and tools used in NeuroIS research

Method/tool	2011	2012	2013	2014	2015	Forthcoming	Absolute	Relative (%)
Papers (total)	6	9	6	20	17	1	59	100
EEG	0	2	0	8	1	0	11	19
fMRI	1	0	1	5	0	0	7	12
SC, SCR, GSC	1	0	0	1	2	0	4	7
ET	1	0	1	0	0	1	3	5
HR	0	0	0	1	1	0	2	3
fEMG	0	0	0	0	2	0	2	3
Mouse cursor movement	0	0	1	0	0	0	1	2
Hormones (saliva)	0	0	0	0	1	0	1	2
Miscellaneous	3	7	3	6	12	0	31	53

Miscellaneous includes both conceptual and methodological paper and the "Relative" category doesn't sum up to 100 as some papers have adopted more than one method and tool

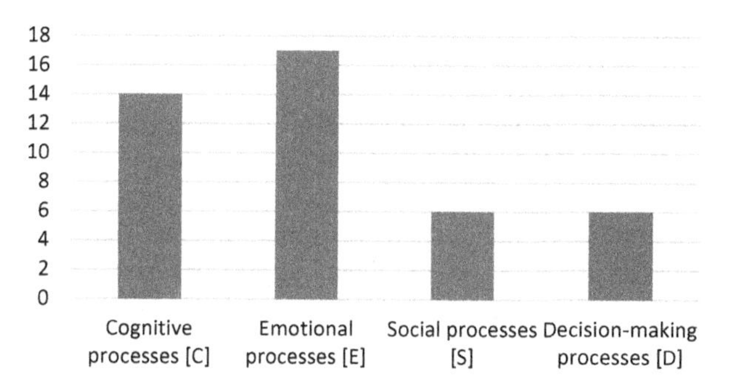

Fig. 1 NeuroIS paper categorization (N = 59) [7–65] (y-axis: absolute no. of papers)

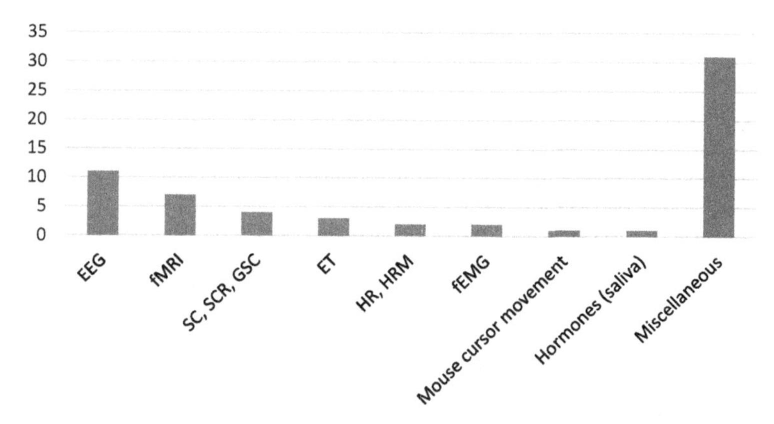

Fig. 2 Popularity of methods and tools in NeuroIS research

4 Discussion

The articles published in above mentioned outlets pertain to significant contributions in different kinds of mental processes like cognitive, emotion, social and decision processes [6,7]. An insight into the overall progression is presented drawing on the results mentioned above, and following implications could be made from each.

A consistent rise has been observed in the past 5 years in the number of publications within the NeuroIS domain as far as the emotional processes and their impact is concerned. Whereas cognitive processes stands second in place. A dearth has been observed in exploration of social and decision-making processes and hence calls for more contributions in these areas. On the basis of our sample (N = 59) [7–65], Fig. 1 shows that EEG is the most popular tool in NeuroIS studies followed by fMRI and eye-tracking.

5 Conclusion

This literature review indicated the importance of neurosciences in IS and how the field of NeuroIS is establishing its grounds in the research community thereby proving its mark as an established field of research. It reported the progress in the field in terms of number of publications pertaining to cognitive, emotional, social and decision-making processes, thereby gauging its popularity. As an outlook, we see that more NeuroIS contributions are needed in order to report studies involving social and decision-making processes. Thus, the IS community can benefit from incorporating the neurosciences tools, methods and techniques by exploiting its inter-disciplinary nature.

References

1. Hubert, M., Linzmajer, M., Hubert, M.: Neural evidence of uncertainty and risk processing networks in information system research: A multilevel-mediation approach. In: Proceedings Gmunden Retreat on NeuroIS. Gmunden, Austria
2. Xu, Q., Riedl, R.: Understanding online payment method choice: An eye-tracking study. In: Proceedings of the 32nd International Conference on Information Systems (ICIS). Shanghai, China (2011)
3. Jannach, D., Zanker, M., Ge, M., Gröning, M.: Recommender Systems in Computer Science and Information Systems—A Landscape of Research, pp. 76–87. Springer, Berlin (2012)
4. Dimoka, A., Banker, R.D., Benbasat, I., Davis, F.D., Dennis, A.R., Gefen, D., et al.: On the use of neurophysiological tools in information systems research: developing a research agenda for NeuroIS. MIS Q. **36**(3), 679–702 (2012)
5. VomBrocke, J., Simons, A., Niehaves, B., Riemer, K., Plattfaut, R., Cleven, A.: Reconstructing the giant: On the importance of rigour in documenting the literature search process. ECIS **9**, 2206–2217 (2009)
6. Riedl, R., Léger, P.M.: Fundamentals of NeuroIS: Information Systems and the Brain. Springer, Berlin (2016)
7. Dimoka, A., Pavlou, P.A., Davis, F.D.: Research commentary-NeuroIS: The potential of cognitive neuroscience for information systems research. Inform. Syst. Res. **22**(4), 687–702 (2011)
8. Anderson, B.B., et al.: How users perceive and respond to security messages: a NeuroIS research agenda and empirical study. Eur. J. Inf. Syst. **25**, 1–27 (2015). doi:10.1057/ejis.2015.21
9. Belanger, F., Xu, H.: The role of information systems research in shaping the future of information privacy. Inform. Syst. J. **25**(6), 573–578 (2015)
10. Davern, M., Shaft, T., Te'eni, D.: Cognition matters: Enduring questions in cognitive IS research. J. Assoc. Inform. Syst. **13**(4), 273 (2012)
11. Riedl, R., Davis, F.D., Hevner, A.R.: Towards a NeuroIS research methodology: Intensifying the discussion on methods, tools, and measurement. J. Assoc. Inform. Syst. **15**(10), (2014). Article 4
12. Tams, S., Hill, K., Ortiz de Guinea, A., Thatcher, J., Grover, V.: NeuroIS—Alternative or complement to existing methods? Illustrating the holistic effects of neuroscience and self-reported data in the context of technostress research. J. Assoc. Inform. Syst. **15**(10), (2014). Article 1

13. Vance, A., Anderson, B.B., Kirwan, C.B., Eargle, D.: Using measures of risk perception to predict information security behavior: Insights from electroencephalography (EEG). J. Assoc. Inform. Syst. **15**(10), (2014). Article 2
14. Léger, P.-M., Sénecal, S., Courtemanche, F., Ortiz de Guinea, A., Titah, R., Fredette, M., Labonte-LeMoyne, É.: Precision is in the eye of the beholder: Application of eye fixation-related potentials to information systems research. J. Assoc. Inform. Syst. **15**(10), (2014). Article 3
15. Teubner, T., Adam, M., Riordan, R.: The impact of computerized agents on immediate emotions, overall arousal and bidding behavior in electronic auctions. J. Assoc. Inform. Syst. **16**(10), (2015). Article 2
16. Astor, P.J., Adam, M.T., Jerčić, P., Schaaff, K., Weinhardt, C.: Integrating biosignals into information systems: A NeuroIS tool for improving emotion regulation. J. Manag. Inform. Syst. **30**(3), 247–278 (2013)
17. Gregor, S., Lin, A.C., Gedeon, T., Riaz, A., Zhu, D.: Neuroscience and a nomological network for the understanding and assessment of emotions in information systems research. J. Manag. Inform. Syst. **30**(4), 13–48 (2014)
18. Minas, R.K., Potter, R.F., Dennis, A.R., Bartelt, V., Bae, S.: Putting on the thinking cap: Using NeuroIS to understand information processing biases in virtual teams. J. Manag. Inform. Syst. **30**(4), 49–82 (2014)
19. Riedl, R., Mohr, P.N., Kenning, P.H., Davis, F.D., Heekeren, H.R.: Trusting humans and avatars: A brain imaging study based on evolution theory. J. Manag. Inform. Syst. **30**(4), 83–114 (2014)
20. Li, M., Jiang, Q., Tan, C.H., Wei, K.K.: Enhancing user-game engagement through software gaming elements. J. Manag. Inform. Syst. **30**(4), 115–150 (2014)
21. Kuan, K.K., Zhong, Y., Chau, P.Y.: Informational and normative social influence in group-buying: Evidence from self-reported and EEG data. J. Manag. Inform. Syst. **30**(4), 151–178 (2014)
22. de Guinea, A.O., Titah, R., Léger, P.M.: Explicit and implicit antecedents of users' behavioral beliefs in information systems: A neuropsychological investigation. J. Manag. Inform. Syst. **30** (4), 179–210 (2014)
23. vomBrocke, J., Liang, T.P.: Guidelines for neuroscience studies in information systems research. J. Manag. Inform. Syst. **30**(4), 211–234 (2014)
24. Hu, Q., West, R., Smarandescu, L.: The role of self-control in information security violations: Insights from a cognitive neuroscience perspective. J. Manag. Inform. Syst. **31**(4), 6–48 (2015)
25. Dimoka, A., Banker, R., Benbasat, I., Davis, F., Dennis, A.R., Gefen, D., Gupta, A., Ischebeck, A., Kenning, P., Pavlou, P.A., Muller-Putz, G., Riedl, R., vomBrocke, J., Weber, B.: On the use of neurophysiological tools in IS research: Developing a research agenda for NeuroIS. MIS Q. **36**(3), 679–702 (2012)
26. Dimoka, A.: How to conduct a functional magnetic resonance (fMRI) study in social science research. MIS Q. **36**(3), 811–840 (2012)
27. Goes, P.B.: Editor's comments: Information systems research and behavioral economics. MIS Q. **37**(3), 3–8 (2013)
28. Adam, M., Gamer, M., Krämer, J., Weinhardt, C.: Measuring emotions in electronic markets. In: ICIS 2011 Proceedings. Paper 1, 5 December 2011
29. Xu, Q., Riedl, R.: Understanding online payment method choice: An eye-tracking study. In: ICIS 2011 Proceedings. Paper 18, 6 December 2011
30. Riedl, R., Mohr, P., Kenning, P., Davis, F., Heekeren, H.: Trusting humans and avatars: Behavioral and neural evidence. In: ICIS 2011 Proceedings. Paper 7, 6 December 2011
31. Senecal, S., Léger P.-M., Fredette, M., Riedl, R.: Consumers' online cognitive scripts: a neurophysiological approach. In: Proceedings of the 2012 International Conference on Information Systems (ICIS2012), December 16–19 (2012)
32. Hubert, M., Linzmajer, M., Riedl, R., Kenning, P., Hubert, M.: Introducing connectivity analysis to NeuroIS research. In: Proceedings of the 33rd International Conference on Information Systems (ICIS) (2012)

33. Zhang, S., Adam, M., Weinhardt, C.: Humans versus agents: Competition in financial markets of the 21st Century. In: Proceedings of the 33rd International Conference on Information Systems (ICIS) [Research-in-progress paper] (2012)
34. Koller, M., Walla, P.: Measuring affective information processing in information systems and consumer research—Introducing startle reflex modulation. In: ICIS Proceedings, Breakthrough ideas, full paper in conference proceedings, Orlando (2012)
35. Grimes, G.M., Jenkins, J.L., Valacich, J.S.: Exploring the effect of arousal and valence on mouse interaction. In: International Conference on Information Systems, Milan, Italy, 15–18 December 2013
36. Kopton, I., Sommer, J., Winkelmann, A., Riedl, R., Kenning, P.: Users' trust building processes during their initial connecting behavior in social networks: Behavioral and neural evidence. In: International Conference on Information Systems, Milan, Italy, 15–18 December 2013
37. Eckhardt, A., Maier, C., Hsieh, J.J., Chuk, T., Chan, A., Hsiao, J., Buettner, R.: Objective measures of IS usage behavior under conditions of experience and pressure using eye fixation data. In: International Conference on Information Systems, Milan, Italy, 15–18 December 2013
38. Anderson, B., Vance, A., Kirwan, B. et al.: Users aren't (necessarily) lazy: using NeuroIS to explain habituation to security warnings. In: Proceedings of ICIS 2014, AIS (2014)
39. Hubert, M., Hubert, M., Riedl, R., Kenning, P.: How consumer impulsiveness moderates online trustworthiness evaluations: neurophysiological insights. In: Proceedings of the 35th International Conference on Information Systems, Auckland (2014)
40. Ortiz de Guinea, A., Webster, J.: Overcoming variance and process distinctions in information systems research. In: Proceedings of the 35th International Conference on Information Systems, Auckland (2014)
41. Bhandari, U., Chang, K.: Role of emotions and aesthetics in ICT usage for underserved communities: a NeuroIS investigation. In: Proceedings of the 35th International Conference of Information Systems (ICIS) (2014)
42. Gleasure, R.: Using distractor images in web design to increase content familiarity: a NeuroIS perspective. In: Proceedings of the 35th International Conference of Information Systems (ICIS) (2014)
43. Neben, T., Schneider C.: Ad intrusiveness, loss of control, and stress: a psychophysiological study. In: Proceedings of the 36th International Conference of Information Systems (ICIS) (2015)
44. Zhang, S., Milic, N.: Carryover effects of system-unrelated emotions on adoption of information systems. In: Proceedings of the 36th International Conference of Information Systems (ICIS) (2015)
45. Neben, T.: A model of defensive information avoidance in information systems use. In: Proceedings of the 36th International Conference of Information Systems (ICIS) (2015)
46. Darban, M.: Emotional foundations of individual's perception: the case of technology radicalness. In: Proceedings of the 36th International Conference of Information Systems (ICIS) (2015)
47. Riedl, R., Rueckel, D.: Historical development of research methods in the information systems discipline. In: AMCIS 2011 Proceedings—All Submissions. Paper 28 (2011)
48. Zhang, Z., Hock-Hai T.: An fMRI-based NeuroIS research on metacognition in the context of e-commerce. In: Proceedings of the 20th Americas Conference of Information Systems (AMCIS) (2014)
49. Walterbusch, M., Gräuler, M., Teuteberg, F.: How trust is defined: a qualitative and quantitative analysis of scientific literature. In: Proceedings of the 20th Americas Conference of Information Systems (AMCIS) (2014)
50. Mohammed, Z., Tejay, G.: The role of cognitive disposition in deconstructing the privacy paradox: a neuroscience study. In Proceedings of the 20th Americas Conference of Information Systems (AMCIS) (2015)

51. Anderson, B., et al.: The impact of technostress on users' responses to security warnings a neurosIS study. In: Proceedings of the 20th Americas Conference of Information Systems (AMCIS), Paper 37 (2015)
52. Ndicu, M., Vedadi, A.: The human-information-processor model view of e-government. In: Proceedings of the 20th Americas Conference of Information Systems (AMCIS), Paper 7 (2015)
53. Barros, V., Ramos, I.: Using Social Media as Organizational Memory Consolidation Mechanism According to Attention Based View Theory (2015)
54. Buettner, R.: Towards a New Personal Information Technology Acceptance Model: Conceptualization and Empirical Evidence from a Bring Your Own Device Dataset (2015)
55. Tams, S.: Challenges in Technostress Research: Guiding Future Work (2015)
56. Bhandari, U., et al.: Follow your heart or mind? measuring neurophysiological responses and subjective judgments for visual aesthetics. In: Proceedings of the 20th Americas Conference of Information Systems (AMCIS), Paper 3 (2015)
57. Zhang, Z., Teo, H.H.: Metacognition in B2C E-commerce: A cognitive neuroscience perspective. In: PACIS, p. 285 (2014)
58. Moridis, C.N., Terzis, V., Economides, A.A., Karlovasitou, A., Karabatakis, V.E.: Integrating TAM with EEG frontal asymmetry. In: MCIS 2012 Proceedings. Paper 5 (2012)
59. Astor, P., Adam, M., Jähnig, C., Seifert, S.: Measuring regret: Emotional aspects of auction design. In: ECIS 2011 Proceedings. Paper 88 (2011)
60. Gleasure, R., Feller, J., O'Flaherty, B.: Human-centred design: Existing approaches and a future research agenda. In: ECIS 2012 Proceedings. Paper 105 (2012)
61. Gimpel, H., Adam, M.T.P., Teubner, T.: Emotion regulation in management: Harnessing the potential of NeuroIs tools. In: ECIS 2013 Research in Progress. Paper 3 (2013)
62. Hariharan, A., Adam, M.T.P., Fuong, K.: Towards understanding the interplay of cognitive demand and arousal in auction bidding. In: Proceedings of the 22nd European Conference of Information Systems—ECIS 2014, pp. 1–10, AIS (2014)
63. Lux, E., Hawlitschek, F., Adam, M.T.P., Pfeiffer, J.: Using live biofeedback for decision support: Investigating influences of emotion regulation in financial decision making. In: ECIS 2015 Research-in-Progress Papers. Paper 50 (2015)
64. Kowatsch, T., Wahle, F., Filler, A., Kehr, F., Volland, D., Haug, S., Jenny, G.J., Bauer, G., Fleisch, E.: Towards short-term detection of job strain in knowledge workers with a minimal-invasive information system service: Theoretical foundation and experimental design. In: ECIS 2015 Research-in-Progress Papers. Paper 24 (2015)
65. Gimpel, H., Regal, C., Schmidt, M.: myStress: Unobtrusive smartphone-based stress detection. In: ECIS 2015 Research-in-Progress Papers. Paper 16 (2015)

Printed in the United States
By Bookmasters